"Educators everywhere are sure to be inspired by the meaningful nature-based lessons Amy Butler and the ECO teachers have created for young children. Their experiences show the power of connection to place, community, and the greater world in learning."

—Rachel Larimore, Chief Visionary of Samara Early Learning

"Amy Butler's book is full of the magic dust you've been yearning to sprinkle on your teaching. The twenty-five lessons are standards connected, introduced in a context of safety and routine, and they are the kind of engaging, curious, active, and creative prompts that ignite a healthy and happy childhood. I found new inspiration throughout Amy's book. I'm off to write some scrolls from Ms. Flora Fauna right now!"

—Eliza Minnucci, author of *A Forest Days Handbook*

"Amy Butler's captivating storytelling narrative in *Educating Children Outdoors* makes readers feel like they are observing an ECO day with her in person. Along the way, you will be inspired by the stories while learning strategies, routines, and lessons that any educator can implement to create successful outdoor learning experiences for students. Everyone should have the opportunity to learn from an educator like Amy, whose passion and expertise shines through in her writing."

—Anna Dutke, Nature Preschool Teacher at Oak Ridge Elementary, Eagan, MN

"One part practical, one part mystical, and all parts beautiful: the words and art of *Educating Children Outdoors* comprise more than just a how-to guide for teaching nature connected curriculum. The reciprocal gathering of ideas are simultaneously a heartbeat, a call to action, and a return to a restful place among the family of things. Educators will be in deep gratitude for this text—a foundational book that has been needed in the field for a long time."

—Anne M. Stires, M.Ed.

"This wondrous book has quickly become an essential resource when I'm teaching children about nature, even though I'm half a world away from where it was written. The knowledge shared by Amy Butler, gathered and learned from many years educating children outdoors is exceptionally welcome to read. The experiential and adaptable lesson plans contained within offer educators meaningful and substantial examples of the hows, whys, and ways to include learning with nature into their education settings."

—Anthony Morris, Director and Early Childhood Educator,
North Fitzroy Child Care Co-operative, Melbourne, Australia

"This book is an inspirational collection of teachers' wisdom and experience of learning and teaching outdoors developed from years of observing children exploring and playing in nature. Their narratives are thought-provoking and provide an empathetic context for a series of lessons in each chapter that demonstrate the huge possibilities of integrating your local outdoor space into our teaching practice. Amy's gentle and eloquent writing style will draw you into the book and ultimately outdoors!"

—Juliet Robertson, author of *Messy Maths* and *Dirty Teaching*

"The beautiful art and inspirational photographs are just the beginning of how this book inspires both new and experienced teachers. Amy Butler and her fellow teachers generously share their stories of growth and success in the outdoors and, in doing so, provide a wealth of knowledge that others can draw from. This is an essential guide to seeing and exploring the nature you have near you. Let it spark creativity and wonder for children."

—Shelia Williams Ridge, Codirector of the
Child Development Laboratory School, University of Minnesota

"Not only does Amy provide practical guidance for those who want to educate children outdoors, she also demonstrates how nature-based learning is interdisciplinary, multisensory, inclusive, and accessible. Decades of research demonstrates that children benefit socially, emotionally, physically, and academically from learning outside. In this book, Amy lucidly illustrates best practices for tapping into all that the outdoors offers."

—Scott Morrison, Associate Professor of Education, Elon University

"A treasure trove of conscientious wonderfulness. If you're an educator hoping to move more of your lessons outdoors, this book may be your best resource. It is the epitome of practical—from behavior management to detailed curriculum activities, it articulates everything you should know and do."

— David Sobel, Antioch University New England, author of *Childhood and Nature*

Educating Children
OUTDOORS

Educating Children
OUTDOORS
Lessons in Nature-Based Learning

AMY BUTLER

Foreword by Cheryl Charles

Comstock Publishing Associates
an imprint of
Cornell University Press
Ithaca and London

First published 2024 by Cornell University Press

Printed in China

Design and composition by
Chris Crochetière, BW&A Books, Inc.

Library of Congress Cataloging-
  in-Publication Data
Names: Butler, Amy, 1974– author. |
  Charles, Cheryl, writer of foreword.
Title: Educating children outdoors : lessons
  in nature-based learning / Amy Butler.
Description: Ithaca : Cornell University Press,
  2024. | Includes bibliographical references.
Identifiers: LCCN 2023006797 (print) |
  LCCN 2023006798 (ebook) |
  ISBN 9781501772047 (hardcover) |
  ISBN 9781501771903 (paperback) |
  ISBN 9781501771910 (pdf) |
  ISBN 9781501771927 (epub)
Subjects: LCSH: Place-based education. |
  Outdoor education. | Nature study.
Classification: LCC LC239 .B88 2024
  (print) | LCC LC239 (ebook) |
  DDC 371.3/84—dc23/eng/20230313
LC record available at https://lccn.loc.gov
  /2023006797
LC ebook record available at https://lccn.loc
  .gov/2023006798

For the children
May you find
awe and peace
in the natural world

Contents

Unit 5. What Does Spring Bring?   187

Afterword   213

Introduction

Teach Who You Are

Journal entry, April 2019

I received a tackle hug first thing this morning as I stepped into the classroom. The first grader beamed up at me and said, "Ms. Coyote! I'm so glad you're here! We have ECO today!"

I'll admit, I get lots of hugs, but this one was a first from this child. I was surprised and curious. As we settled into morning meeting with both classes, thirty children in total, the energy was a bit damp. Not the usual for our weekly Wednesday mornings.

The teacher whispered to me, "We have some children needing a little extra care this morning." Maybe it's because it's April, the sun is hiding, and the clouds have formed a gray roof over our sky. Or maybe it's because children are people too; life is hard and some days we start with frustration and tears.

"It's a perfect day to get into the forest," I whispered back.

The children I work with know me as Ms. Coyote or Amy Coyote. In the first few weeks together, spending time outdoors and getting to know one another, they tip their heads sideways and look me in the eye, asking, "Is Coyote *really* your name or just a pretend name?" I guess it's real and pretend. Just as the students choose a nature name that resonates with them, and which becomes an animal they get to know with research, I have always identified with the coyote, and so it has become my nature name. Coyotes have always intrigued me. Related to wolves and a distant cousin of our beloved dogs, coyotes are between worlds of domestication and being wild. It's a bit like how I feel in my role as an educator.

By the time they have all chosen their own special nature names, the questions about who I am and where I live and if I am young or old fade away. At that point, we are deep into our weekly routines of learning outdoors, and the questions have changed as our relationships with one another have become stronger. The children, the teachers, and I are now on "nature time," which affords us a different way to build these connections with one another. Our nature names seem to always be a wonderful catalyst for making these connections.

These names are enduring. Some children, who are now sophomores and juniors in high school, I know only by their nature names. Spring Peeper and Jumping Mouse both come to the forefront of my mind. These two children, now teenagers, were some of the many who affirmed my thinking about how every child can benefit from time immersed in nature with their peers and caring adults. The data on the benefits of time spent outdoors, which are only an online search away, were proven when Spring Peeper found her voice each week in the forest. Having not spoken at all in the classroom, and with a possible diagnosis of being selectively mute, she did speak on our forest days, we noticed. It happened slowly at first, by accepting help from Jumping Mouse to climb a steep hill; and then again, later, when she joined in the imaginative and meaningful play of building mouse houses.

Communication is essential when we are outdoors learning and exploring, and it can come in many forms. The same way nature communicates with us through wind, waves lapping on the shore, or two branches clacking together, children outdoors in nature figure out how to communicate with each other with and without words, even at times without the same spoken language. Most days, Jumping Mouse was ready at Spring Peeper's side, encouraging her play and interaction with others. The child-centered and intense physical nature-based play, along with the natural environment, supported new relationships for these two children, through which they both learned to communicate with their peers. This experience could not be replicated within the four walls of the classroom.

Jumping Mouse helped Spring Peeper build a bridge into the forming social ecology of the kindergarten class in the forest. Forming new relationships outdoors was helping Jumping Mouse too: Where he struggled inside the building in peer-to-peer relationships, he shined in the forest by sharing his familiarity with the outdoors and gifting that to others in acts of caring and kindness. As for Spring Peeper, we soon noticed her growing comfort outdoors and her increased verbal engagement—not only with her peers, but with her teachers, too. We documented these moments and quietly supported the children, allowing our time in nature to be the wisest teacher at that moment.

For many of us who dedicate our careers to teaching, we do it because we care about the future of our youth. We become teachers because we want the educational system to be different and better than what we experienced. And for some, we become teachers as a form of activism. As I started my path toward being a classroom teacher, I knew I could better serve my students and be a happier teacher if I could support children in the way I learn best. For me, that means I need to have choice, autonomy, lots of movement, purpose, and as much time outdoors as possible. When I was an elementary student in the early 1980s I silently struggled through all academic tasks. I kept my head down, behaved, and filled in the blanks. It wasn't until I was in high school, when I failed chemistry and physics twice, that I was offered an alternative: I could join Future Farmers of America (FFA) and get my credits for graduation that way. Outdoors, tapping maple trees, tilling fields, and measuring the diameter of conifers, I finally found joy in learning. It was the hands-on approach that fired up my brain and ignited my creativity. I felt calm and capable. I discovered who I was as a learner in what a public school system called a "nontraditional" setting. See, kids like

Chapter 1
Nature-Based Routines
for Outdoor Learning

If the love of nature is in the teacher's heart, there is no danger:
such a teacher, no matter by what the method, takes the child
gently by the hand and walks with him in paths that lead to the
seeing and comprehending of what he may find beneath his feet or
above his head. And these paths, whether they lead among the low-
liest of plants, whether to the stars, finally converge and bring
the wanderer to that serene peace and hopeful faith that is
the sure inheritance of all those who realize fully that
they are working units of this wonderful universe.

—*Anna Botsford Comstock,*
Handbook of Nature Study, *1911*

A Morning at ECO

Narrative by Amy Butler

Like a murmuration of birds, the children run and skip through the open soccer field to our morning meeting place. This active burst of flight was just preceded by a handful of minutes of the students silently observing the morning landscape right outside the school building. In this transition from the building to the outdoors, children and teachers observe the clouds, gather weather data, and listen to the soundscape for birds. I notice icy dew on the grass and share my find with a student. One student's job for the day is to be the photographer, and they stand in a marked spot and take a picture of the schoolyard tree. There is a collective and purposeful way in which the students and teachers move into the outdoor world. This is not the typical exit from the building for recess; today is an ECO Day, and our routines for learning together outdoors are in motion.

After landing at our meeting place on the far end of the soccer field, some children begin to form a circle and are singing a song they know well: "Good morning, Earth! Good morning, sky! Good morning, water flowing by. . . ." The song cues everyone to come together, and the circle continues to grow. Some children on the outskirts are still in tumble mode, and I gather them up and bring them into the formation. Our morning meeting is about to begin, and bodies settle into mountain pose, ready to share. As our song comes to a close, children's hands shoot up, and our round of nature notes begins. Children are very eager to share what they have noticed in nature this week. A white-tailed deer in their yard, a spider in their bedroom, and the first red leaf of autumn. As the sharing moves around the circle, bodies begin to wiggle. I can see the energy and focus has shifted, and it is undoubtedly time for a game!

Our game for the morning is based on a predator and prey relationship of owls and mice. The students' fascination for a resident Barred Owl has sparked an exploration of owl species in our region, and the predator-prey relationship has

Nature notes

A nature note is a shared observation of something a person has noticed in nature. Nature notes may be about the weather, animal sightings, plants or trees, or natural phenomena. Sharing nature notes on a weekly basis brings elements of the natural world immediately into your teaching practice. A nature note is something everyone can access, because nature is everywhere! These observations can be gleaned from a cityscape or while waiting for the school bus. This is an opportunity for students to practice oral language development. It also lays a rich foundation for future storytelling. Nature notes can be recorded throughout an entire school year. What has changed seasonally? What are the common nature note themes? This is an opportunity to practice phenology: the study of cyclic and seasonal natural phenomena.

Amy Butler

been scaffolded in the classroom through reading nonfiction text. The children are primed and ready for our game of owls and mice because they now have prior knowledge of this species. Every outing, we play a game before we move into the forest or urban park to our outdoor classroom. This is a way for students to release energy, get warmed up, and connect as a classroom community. Play is a fabulous facilitator of development in social-emotional skills, and pretending to be animals is a fun catalyst of play behavior. This routine of playing games allows students to practice motor skills and apply some basic science concepts through creative movement. And it also buys us some time as teachers to get everyone out the door, accounted for, and

settled into the beginning routines of our morning before we depart any further from the building.

After a few rounds of Owls and Mice, children are signaled to gather their backpacks with the sound of a chime. Once again, I can hear a pack of children begin to sing unprompted! The two classes hustle to the edge of the forest, ready to hike up to our basecamp. Within this transitional routine we have safety procedures so embedded that this organized chaos falls into line surprisingly quickly. A safety sandwich is formed by the adults, and students are counted. Our magic number of children for the day is twenty-five.

Bread is established and sandwich fillings are now prepared. Today I decide to make two safety sandwiches, because I know there is a front group of runners eager to explore, and another group that relishes the idle morning chatter found on the trail. It is just a good day to meander and soak up the benefits that the fresh air and open space give us. As the children and adults enter the forest, some slip on their quiet fox-walking feet. I notice students balancing on one leg and reaching down to put on invisible socks. This fox-walking routine was introduced last year in kindergarten, and the children still practice this threshold ceremony as they enter this wild space. A few children greet the forest with a "Hello!" and "Thank you." Our morning has just only begun.

Teaching and Learning with Nature

The previous narrative is a glimpse of an ECO session that could be happening at any public school. Efforts to get students outdoors on a weekly basis have continued to sprout in elementary schools all over the country. From Forest Fridays to Woods Wednesdays and, yes, even Terrific Tuesdays, the evidence of children finding new paths of learning through forests, fields, and in city parks is evident. Teachers and parents have wanted for something different, and nature-based programming is delivering something new. As word has spread, a collective understanding of the benefits of this time outdoors has grown. Kids love the days they have ECO; it has become a favorite day of the week.

With the growth of nature-based programming across the United States also comes a plethora of resources out

ECO basecamp

A designated undomesticated and somewhat wild space in nature that serves as a gathering place for learning and exploration. The basecamps can be on school property or public use land. These sites are mitigated for hazards (see appendix 4), surveyed for their plant and animal species (see appendix 3), and adopted for use by schools. These spaces are visited on a weekly basis and serve as an outdoor classroom.

Safety sandwich

A safety sandwich is a simple safety protocol for traveling from point A to point B with a group of children. One adult (and student helper) is at the front of the line as a "piece of bread" and another adult (and student helper) is at the back of the line as the other piece of bread. The children are the filling of the sandwich. Be sure your sandwich doesn't lose any filling and ooze out! The pieces of bread are responsible for keeping everyone together. Sometimes it's helpful to have two sandwiches!

Magic number

The magic number is another built-in safety protocol that involves the students in creating a habit of counting students and being aware of who is here today. There are lots of ways to count the magic number, depending on students' ages. Sometimes we are in a rush, and it is easier to count them ourselves. When I don't include the children, someone always reminds me: "Stop! We haven't counted the magic number yet!" The magic number is counted during transitions—after games and activities when children have been spread out, at a trail or street junction during travel, and always before leaving the school grounds and before leaving an outdoor setting. Designate a student to be the magic number holder of the day and have them count or confirm the count at these check-in times during the session. The Magic Number practice was inspired by Juliet Robertson, outdoor play advocate and teacher.

there for teachers, from books to blog posts to hashtags and Facebook groups to podcasts and oodles of flashy Pinterest pages. It's a lot to navigate, and finding ways to make it work for your class can be daunting. Common concerns start to emerge in the form of questions: How do we bring students safely outdoors? Will they be focused enough? What about the curriculum? And then sometimes there's this one: I don't know anything about nature!

All these questions are important. What I want teachers to know is that teaching and learning outdoors, every week with nature, is possible. How this practice looks here in Vermont will be different where you teach. What is socially and culturally relevant here in the rural pockets of the Northeast will be unlike a vibrant cityscape where nature is visible in parks, sidewalks, or playing fields. But here's what I've learned from working with children outdoors: Nature is everywhere. Nature is the urban stream and the plants reclaiming an abandoned lot. Nature is the open sky above our heads. Every child lives near nature, and we can help them connect with it.

The first step out the door into the world of teaching with nature is a big undertaking—and, I will add, very brave. With all the many things that teachers are asked to do, it is one more thing to develop strategies of teaching without four walls and a roof. You'll have to resource the needed clothing, plan new curriculum, advocate for room in the school-day schedule to leave the building, and overcome the barriers to accessing a safe place in nature to learn. My friend and colleague Eliza Minnucci puts it very simply in *A Forest Days Handbook*: "Try it. Just try it. If you find joy in nature, if you find inspiration, wonder, calm, or awe . . . trust that when you bring

Did I have some questions and concerns early on? You bet. My questions included: Am I dressed appropriately? Does my own backpack have everyone's inhalers, meds, etc.? Are all children being properly clothed and booted up? Would I know enough about the subject (science and environment) to be effective with the children? How will parents perceive this time the children are spending outdoors? Will I have the energy to keep up and also be able to support positive behavior? How will this particular group of children handle the behavior expectations of being outside given the independent learning challenges? I could go on and on listing fears and anxieties that would crop up. These questions and concerns could have easily prevented me from moving forward and taking some chances, thereby never allowing the first and second graders to have this experience. It would have been easy to use my fears or the unknowns as excuses to not try. There was a relationship of trust that developed, which I think needs to be part of team teaching, especially in an outdoor environment where risk and hazard needs to be assessed. Each teacher is able to use their strengths with a group of children to help them become leaders of their own learning.

—*Anonymous feedback from a teacher interview, 2012*

drawings, which later developed into independent writing motivated by their time outdoors.

This integration back inside the classroom is what differentiates a field trip to nature from a nature-based experience like ECO. The experiences in nature will come back in with the children, so give plenty of opportunities to expand the learning. We want to weave together the threads of the learning indoors and the journeys into those unbound spaces in nature. This will make for a complete learning cycle.

When the sun comes up again tomorrow, even if you don't plan to go outdoors for any learning, the children will still be holding onto the energy and experience of the day before. Don't let it vanish. In units 1 through 5, you will learn more about how to scaffold the learning process and connect it back into the classroom.

We have adopted these routines and have practiced them in a variety of settings. They are certainly not rigid, and we are conscious of the saturation or stagnation in following these routines. Some days, things just need to be shaken up or released. The mood of the day, the weather, and the welcomed distractions of nature all play into this. The fascinating thing is that, because these routines are continuously practiced over time, they become habits that create a culture of learning that is unique to being in nature.

Because of this, the children could run the time outdoors themselves—and sometimes they do. Like a murmuration of birds, the children are guided by these routines with no adult leading. There are moments when a teacher, or I, find

Amy Butler

ourselves observing the active hum of students engaged in purposeful work and play. We realize no one needs anything, no one wants for anything. The routines we create for learning in an outdoor setting are much more meaningful when we think of them as routines of connection, their purpose being to build stronger, more diverse, and more resilient relations with ourselves and one another.

Teaching outdoors promotes full body learning and full body engagement. Based on my observations over the years, being outdoors promotes a healthy and physical experience in which deep learning occurs versus a more structured and stressful learning experience that mainly requires a lot of seat work. With our regular outings, nature becomes an embedded tool that guides and teaches our students.

—*Roberta Melnick*

Chapter 2
Awareness and Safety as a Daily Practice

Narrative by Amy Butler

It was eight weeks into the school year and the leaves now blanketed the ground in the forest. The plan for the morning was to collect small sticks the size of a woodland jumping mouse's tail. We would use our harvest of tiny twigs for our very first fire. With the deciduous trees being bare it meant our usual area for learning and exploring would be covered in leaves. Not only that, it had been raining heavily all week. This meant the task of gathering was going to be met with some challenges in finding not only sticks but also dry ones that would make for good fuel. Not too far away was a grove of eastern hemlock trees, a favorite fuel for fire. The downed lacey branches of this conifer were easy to gather from the forest floor, and we would be dry as they were sheltered from the rain under the boughs of the trees. Plus, we had never been there before, it was sure to add a sense of adventure to the morning. We quickly decided to change our location for the day, and because the students were so familiar with our routines for learning outdoors, we expected it to be seamless.

After a round of Bear Tag in the mowed playing field, we announced we would walk a bit further to a new area to look for mouse tails. Today we would not be visiting our regular spot in the forest. The students fell into our safety sandwich, and after counting the magic number we made our transition. We arrived at the edge of the hemlock grove in an adjacent field, and we stopped to review what boundaries would look like in this new forested home. As the class discussed safety in this new space, the teacher went into the forest and did a site assessment. They looked for hazards such as hung-up tree branches overhead, any human refuse, and considerations for staying dry.

"You always need to be able to see an adult," shared one student.

"If you see broken glass or something like that, tell an adult first, don't touch it!"

"If someone is hurt, help them. That's the second care. Taking care of others."

"Work with a buddy! And when you hear the crow call come back."

"Yeah, but when you hear the crow call, give a crow call back. So it spreads."

The previous eight weeks seemed to have been effective. The students were confirming their understanding of safety in this new space by replicating the routines we had been practicing in our outdoor classroom. They were adept at recognizing hazards and negotiating risk for themselves. With a thumbs-up from the teacher, who assessed this postage-sized slice of forest, we embarked on the task ahead.

Students partnered up with one another and then fox-walked, bounded, and pranced into the hemlock grove. The adults spread out as boundary posts, and the sounds of children oohing and ahhing under this new species of trees could be heard as they made discoveries.

All except one student.

On the edge of the forest standing in our meeting place was one child. They stared into the forest with their arms crossed, feet planted firmly on the ground. I was curious about what was going on and stayed where I was, observing. The student started to turn and walk away and then abruptly stopped at the sound of shouting friends. They returned to their previous post. The classroom teacher saw me and gave me a wave. I nodded back as to signal "I'm here, I've got this," and they gave me the thumbs-up. On this day we had a total of twenty children, one classroom teacher, one paraeducator, and two parent volunteers, plus myself. A twenty-to-five ratio of children to adults. At any point the ratio could fluctuate between four-to-one and five-to-one, depending on the needs of students. Right *now* was one of those times.

I slowly walked closer to the student who now had her arms stiff at her sides and was kicking the ground.

"Hey," I half whispered. "Can I help?"

She didn't look up at me but very sternly answered back, "I'm not going into that stupid forest. No way."

"Okay. I can see you don't want to. Do you feel like talking about why you don't want to?"

"That game is stupid. I'm not doing it," she tells me.

"Oh." I allow for a long pause. "On our walk today you told me you were excited to collect mouse tails. Do you want to talk about what changed? I am here to help." At this point I was standing shoulder to shoulder with the student, and I could hear she was breathing irregularly and holding her breath a bit. As she watched her classmates, her eyes darted back and forth. She was obviously uncomfortable and processing something big. I took a deep audible breath and slowed my own breathing. Co-regulation felt really important in this moment. Having a trusting relationship with this student enabled me to be close, ask questions, and take things slowly.

"My mom said to not go into those kinds of woods."

My mind began to scramble. I stayed close and decided to kneel on the ground; I got out my water bottle. I needed a moment to pause and get a sip of water. My heart was racing as I pieced together what I knew about this child's life experience in her short six years.

We had been in the forest for the past two months and this child was flourishing in a way she had not been back in the classroom. The teacher was seeing another side of this child that presented as kind and brave. What changed? What was different? Change in a setting where a child has felt safe and comfortable is difficult. And children can come to school on any given day with the weight of the adult world in their pockets. That weight, whether it be a small sharp rock or a heavy granite stone, will dictate the day ahead. This seemed to be one of those days for this student.

We were at a peak of the opioid epidemic in Vermont. I knew enough about the struggles in this particular community that I sensed what was going on for my friend. Discarded needles had just started to show up in some counties, primarily in public use spaces and parks. There had recently been a spike in crime and arrests as well.

I stayed shoulder to shoulder with her and offered that we could sit right here and together we would watch until the activity was done. The student agreed and we settled into drinking water and getting out our snacks. I reassured her that we could see everyone, even her teacher and the other adults. That's what we do on ECO days. We take care of each other. And right now, we were taking care of ourselves, too, by listening to our bodies. We continued to rest, watch, and breathe. Her shoulders relaxed and she leaned against me. We began some idle chat that turned into a request to get out my bird field guide because she thought she saw "the little camo brown bird that spirals up the tree."

The rest of the class continued to dash around in the forest. A few classmates noticed us sitting in the browning tall grasses on the edge of the forest. They came over to inquire what we were doing and asked if they could join us. I waited for my friend to answer first. She said yes. "We are looking at Amy Coyote's nature books. What else do you have in that backpack anyway?"

With a wave and a whistle to the classroom teacher I signaled that I now had three more students. We were back to a four-to-one ratio. The first of the three cares was being modeled, care for self, and the new trio followed suit. The children settled into a pack and proceeded to get sips of water and a snack out of their backpacks. The conversation of the now four children shifted from the field guides to the mouse tails being collected. There was some affective sharing happening with oohing over the bird guide and pictures of the Brown Creeper. One student had let go of their fist-sized handful of mouse tails and placed it where everyone could see. The child who had not entered the forest eyed them and commented, "Those look like good tails. Where'd you get them?"

"Right there under that tree," pointed the child. "They are all over the place. Wanna see?"

This is when the second care came swooping in, caring for others, as these two students connected over the excitement of mouse tail–sized sticks. With no plan or prompt from me as the adult, they both jumped up and ran to the tree to search for more.

My friend who had been on the edge was now absorbed by her peers with a renewed sense of self and belonging that felt safe.

We had about five more minutes left of our time to collect sticks. I decided to observe within earshot and see if we might stay a bit longer. The sun popped out for a moment, and I felt the warmth relax my body and gave one more deep exhale from my belly.

Caring as an Act of Safety

> The courage to teach is the courage to keep one's heart open in those very moments when the heart is asked to hold more than it is able so that teacher and students and subject can be woven into the fabric of community that learning, and living, require.
>
> —*Parker J. Palmer,*
> *author, educator, and activist*

Parker Palmer's words remind me of what we experience as teachers when we venture past the four walls of our classrooms: that to teach without walls takes courage and an open heart. It also requires a committed school community. In the preceding narrative of a morning in the forest, we learned about a student who was afraid to enter a new forested area and how the student was supported in the outdoors, with no walls. Sprinkled in the narrative are clues to safety practices that help to keep students safe and

Amy Butler

the adults aware of the needs of children, including the safety sandwich, counting the magic number, knowledge of how to mitigate hazards, and a child-to-adult ratio that supports safe and meaningful learning. These are all protocols that have been developed into routines when we are outdoors. Beyond the logistics of these safety measures, we need to also be present and caring with students in the most challenging moments. As teachers we are responsible to meet our students where they are with new experiences in the natural world and to keep them safe. Not only physically but also emotionally. The very thought of keeping children safe in an undomesticated and wild space in nature can, at times, be asking our hearts to hold a lot. And we need not only the courage to do so but also the knowledge and a framework that makes getting out the door and into nature accessible for all—including ourselves.

This section will cover safety in the form of the three cares. How can I be safe outdoors by caring for myself? How can we be safe outdoors by caring for others? And how can we take care of the Earth when we are learning in nature? The three cares—caring for self, others, and the Earth—is a brilliant framework that reflects the ethics of care as written by Peter Martin in his paper "Caring for the Environment: Challenges from Notions of Caring" (2007). In it, Martin examines the structure of caring and explains that how we model caring for self, others, and the Earth can have implications for the effectiveness of environmental education. He concludes that how we practice caring for other people can be transferred to caring for the Earth in various ways. Martin

The Three Cares

The three cares act as a framework for building a safe, inclusive, and caring community in a nature-based setting. Students learn to care for themselves, care for others, and care for the Earth. Introducing students to these three cares right from the beginning helps students and teachers develop a common language that guides our habits and behaviors when we are outdoors. The three cares are an extension of classroom and school expectations. How will I care for myself in nature today? How will we care for others? How will we care for the Earth?

The term "The Three Cares" has been inspired by teaching practices at Earthwalk Vermont.

many years teachers have shared with me that practicing these three cares is essential to being outdoors successfully, not only for safety but also for guiding students to develop relationship skills and for making responsible decisions. It is the one teaching tool they wouldn't leave the building without. The ways in which we care for ourselves, others, and the natural world are constantly evolving and growing just as the students do. As the seasons change, so do the three cares. I imagine if there is a safe, inclusive, and accessible place for students to come back to during ECO, it would most likely be in the form of a tree resembling these three tenets. Caring for ourselves is the heartwood of the tree. Caring for others is the branches of the tree that shade us all. Caring for the Earth is the roots of the tree that connect us to place and to the land. Let's start with the first care, caring for ourselves, located in the heartwood of the tree.

highlights that caring is an emotional act. When we care, we feel emotions, and this acts as the glue in relationships. Emotion is also a principle for motivation to act on behalf of others. Another connection between human relatedness and our connection to the natural world can be seen in friendships. Friendships need care, and they thrive with reciprocity. Having friendships is crucial to our human experience, and the skills we learn to make friends and nurture friendships can be applied when connecting to the natural world. In conclusion, Martin also states that by providing students with language (like the three cares) around our human and nature relationships, we can build children's ecological literacy and create new perspectives on our role in caring for the Earth.

Today, the practice of these foundational cares can be seen reflected in the teaching philosophies and practices of many environmental education organizations and nature-based programs. For

Nick Neddo

Caring for Self

Before we take students into nature, we can reflect on our own relationship to nature. Caring for ourselves begins with our very own self-awareness in the outdoors and our ability to manage ourselves in relation to nature. What do you love about being outdoors? Do you relish cold wintery days, or would you rather stomp in rain puddles? How will you manage being wet, dirty, or sweaty and hot? Maybe you prefer leading students outdoors only on warm days when the sun is shining (this is okay!). What excites you about the natural world? As educators we need to be aware of our own edges for being outside for long periods of time with children, and remember what brings us joy. Think about how you want to feel as a teacher each day. Now think about how you want to feel as a teacher outdoors with nature as your co-teacher.

If we are not aware of how to take care of ourselves in nature, we cannot be co-learners with our students. If the adults are unprepared or unaware of hazards, they will be distracted and not able to engage in the joy and curiosity of learning with nature. The child in the above narrative who would not enter the forest out of fear could also very well be the adult, the paraeducator, or the parent volunteer. A teacher's first responsibility is to maintain a safe learning environment for the students, and yes, that means safe for the adults too. Learning to care for ourselves in nature is an extension of how we teach our students to prepare to learn with nature. Below are four areas in which we can consider how to care for ourselves before heading outdoors with children.

Practicing Self-Awareness and Self-Management

Check in with yourself and your coworkers about how everyone is feeling today. Are we mentally ready to face our first rainy day outdoors? Set personal and collective goals for your day outdoors as a teaching team. This practice continues as your students fill the classroom and come to morning meeting. Knowing how students are feeling will help you navigate the time outdoors better. Maybe today a strong review on dressing for cold weather or a reminder on our three cares is needed. Who needs some extra support today? Take a break if you are feeling frustrated or angry. A round of mindfulness breathing exercises can bring our cortisol levels and heart rate down. This is care for self.

Clothing for All Seasons and All Weather

Everyone needs the appropriate clothing and gear for the weather. Clothing is a form of shelter. This shelter for our bodies is also a form of safety. When we are warm, dry, cool, and comfortable we can fully participate. This is true for both the adults and the students! As the teacher you need to be able to sit on the ground, run and play games, and demonstrate to your students how to safely and joyfully engage. Being hesitant to kneel on the ground to speak to a child because you will get wet is a barrier to connecting with your students. We also need to consider students with tactile sensitivities to clothing. How can we support them? A rain hat as opposed to a hood may be a solution for some children. Every child needs shelter in the form of clothing. Lesson 17 will cover dressing, and appendix 8 suggests ways to get the needed gear for your school.

Drinking and Eating

Our brains need to stay hydrated and fed in order to fuel learning outdoors. We know this to be true inside the classroom, now consider being outdoors on a 20-degree winter day or an 80-degree summer day. Water and food are essential in keeping us warm or cool. Drinking and eating are also a part of our routines. Sitting with peers and sipping tea or having snack also develops a sense of belonging. As the teacher, pack a favorite high protein snack and thermos with a warm or iced beverage and bring extra for your teaching partner.

Sleeping and Resting

Much as with food and water, we know our brains work better with plenty of sleep and rest. It is helpful to give a reminder to students and ourselves that the next day is an ECO day, so we all need to get a good night's sleep. After a day of active learning outdoors, our bodies need adequate rest to rebuild and rejuvenate for another day of learning. Sit spot outdoors or a time for rest and reflection back inside the classroom is another way to work in this essential form of self-care.

Amy Butler

Our twenty-first-century lives are largely spent indoors, whether at home, at school, at work, or in a vehicle for transportation. The average American child spends between four and six hours each day with some form of media. We can assume that our students may have little to no experience in an unbound, forested area or the wild spaces of an urban park. Students' previous exposure to the outdoors and knowledge about the natural world are foundational to how you will scaffold your exit from

When you have people who have a certain amount of access to nature and then you give them a bit more, you see better social functioning. You see better psychological functioning and better physical health.

— *Ming Kuo, Associate Professor of Natural Resources & Environmental Sciences, University of Illinois at Urbana* the indoor learning environment to the outdoor one. Again, think of this as the heartwood of the three cares. When caring for ourselves in nature, we can start with asking, Who am I in nature? Or, who are the children in nature? By consulting children first about their experiences in nature, we are letting them know that we value all their diverse backgrounds. This includes understanding the broader social and cultural norms of your classroom community, and it means that we include them when preparing to spend consistent time outdoors. We can begin to consult with children and weave in safety by asking these three questions. These questions are inspired by Tim Gill, an advocate for children's right to play outdoors in safe spaces.

1. What is wonderful about learning outdoors/in nature/in the forest?

   This first question is very intentional as it starts with what is positive about spending time in nature. Here we are getting a baseline for where students are at. What do they like about being in nature or in a forested setting? Have they visited the town forest or city park? Have they spent time fishing or swimming? Do they have a family member who keeps a garden or feeds the birds? You will see minds and faces light up with stories, and some children will be mentally digging for a connection. What children share about what is wonderful is a window into their relationship with nature.

2. What do we need to be careful of (or mindful of) when we are learning outside/in nature/in the forest?

This question is where the children's feet hit the trail, and we listen to what they believe we need to be careful of. What do they perceive as a hazard in a natural setting? Are they concerned about tigers or insects? Do they wonder if they will get lost or trip and fall? You are gathering information that will inform you as to how they might react to and interact with nature. Again, what is socially and culturally relevant here as dangers in their world will come up. Some will be very accurate, and some may not. What is important is that the children's concerns are being heard and that we address them. When consulting children, it is important we respect their thoughts and ideas. This is how we show we care about their thinking and include them in the process. As the teacher you will be mitigating hazards through performing a site assessment (see appendix 4) and creating emergency protocols for your program (see appendix 5). The

Each school that embarks on a weekly nature immersion for students needs to carefully consider what its emergency protocols would be for its school community (for more on this topic, see appendix 5). The ECO Protocols for Safe Outings have been developed after years of trial and error. Some protocols were made after we needed them. Each school is unique in its geographies and also access points to places where teachers and students can partner with nature to learn. And how we navigate risks and hazards and develop safety protocols is entirely dependent on who your students are, their needs, and the specific places in which you plan to learn.

third question involves the children in creating the rules for safety.

3. How will we stay safe and care for each other?

This final question invites children to be active participants in making the rules for being safe outdoors. We are involving them in problem-solving how to mitigate risks and hazards. We are collectively agreeing on how we can care for one another (back to the second care!) in the outdoor classroom. When we ask children about how we can stay safe, we're letting them know that we trust them and that we believe they are capable of making decisions about safety.

When asking these questions, write down exactly what the students say. Keep a documentation of the answers. The class can come back to these three questions in another season or to focus on a specific activity such as stick play or tool use. How do the students' perceptions of safety change after they have had more experience? By using these three questions we are immediately inviting the children to help make the rules and plan for their time outdoors through the lens of safety. The expectation is that we continue our school rules and agreements when we are outdoors. Being kind to one another and caring for each other are also ways we can stay safe.

Caring for Others

When we are able to care for ourselves, we are then able to care for others. Much like the airplane analogy about putting

Amy Butler

on your own oxygen mask before helping others, teachers should put on their own backpacks before heading into nature. If we are feeling rested, hydrated, fed, and sheltered, we are able to be supportive of others around us who need some extra help. Children can list numerous ways in which they practice care and kindness in the classroom: Help people when they are hurt, use kind words, offer a hug or a smile if someone is sad. How does this look and feel different in an outdoor setting? It doesn't look that much different, but because the environment is completely different and offers specific challenges, it brings caring for others to another level. Children will demonstrate compassion and empathy for others in relation to the weather, the landscape, and the activities or forms of risky play in which they are engaging. When another student trips over a root in the forest, is cold, or feels sad over a lost slug, children can feel a real sense of agency in helping another child. As they are helping one another in nature, they are navigating a unique setting where

Everyone's first and favorite question: What about the bathroom?

Toileting is a very important routine for caring for ourselves when we are outdoors. Different situations call for different solutions when it comes to toileting in a nature-based program.

Before you leave

- Who are your students? Consider the time of day and usual high bathroom use. Remind students how long you will be out. And definitely use the bathroom before you leave the building!
- Pack toileting supplies in your backpack (see appendix 6)

Now that you are outdoors

- Can you re-enter the school building for toileting? This depends on your adult-to-child ratio and how far you need to travel. Who in the school building can assist?
- Is there a public bathroom available nearby? Again, ratios matter. One adult will need to leave the main group and assist students.
- "Wild Toileting" may be the only option in an emergency. Learn the leave no trace way for wild toileting. Always have toileting supplies in your backpack. Some pre coaching with students before the time comes goes a long way!

If you are considering creating a site-specific toileting option

- Is a portable toilet an option for your site? Do your research. Portable toilets should be environmentally sustainable and hygienic. How will you dispose of waste, and who gets the honor of carrying it back to school?
- Chemical toilets can be rented and maintained for a school year or seasonally. Location matters. Consider distance from your outdoor classroom and the possibility of public use.
- Some state parks and national parks use a moldering privy for human waste. Is this an option in your community? Who will maintain it?

Most importantly

- Any outdoor toileting must follow procedures for best hygiene practices.
- Follow Leave No Trace protocols (see appendix 13)
- Always respect children's dignity and privacy. Children need to feel safe, supported, and cared for when toileting outdoors. When they do, it becomes less of an unknown and another helpful skill learned for being in nature!

Additional resources

Creative Star Learning Blog Post—Outdoor Toilets and Children by Juliet Robertson https://creativestarlearning.co.uk/early-years-outdoors/where-to-go-when-you-need-to-go/?
Inside Outside Nature-Based Educators—Portable Toilets. https://www.insideoutside.org/portable-toilets

they must communicate effectively and be active listeners. Showing leadership, problem-solving, and evaluating consequences are all skills that are learned in nature. These are also foundational skills to building positive and healthy relationships. These situations and opportunities to care for others cannot be replicated inside the four walls of a classroom. As children become more connected to their environment and more comfortable with the routines of being in nature, their capacity to care for others increases.

Learning how to care for others outdoors is crucial because there are inherent risks and hazards in nature. Children are naturally drawn to play and learn in nature because it has a level of risk. Moving fast, hiding during sneaking games, using real tools, rolling down hills with friends, and balancing and swinging on natural elements are all things you will see children want to do and crave to do. Learning to negotiate and assess risk is an important part of early childhood development, and the natural world has many opportunities for children to

Amy Butler

initiate and experiment with risk-taking on their own. When we allow children to follow their desire to take risks in an outdoor setting, they have a chance to practice self-assessment skills, and we are investing in their development to be able to make thoughtful and responsible decisions in other areas of their lives.

Knowing the difference between a risk and a hazard is foundational to having a safe learning environment outdoors. In order to dynamically assess risk for ourselves and with our students, we must understand the difference between the two. In Natural Start Alliance's publication, *Nature-Based Preschool Professional Practice Guidebook*, risks and hazards are summarized this way:

Risky play and learning are only developmentally appropriate when they are managed and supported by an attentive and caring adult. We can actively involve children in the risk-management process by simply discussing the benefits of a

> Teaching children to assess the benefits and risks of an activity supports their individual agency in the world. They learn to determine for themselves if and/or how they will engage with their world by intentionally weighing the potential upsides and downsides.
> —*Rachel Larimore, author of* Preschool Beyond Walls: Blending Early Childhood Education and Nature-Based Learning

Risks versus hazards

Risks	Hazards
Are easily identifiable	Are difficult or impossible to assess (by the child)
Can yield growth	Can cause harm
Can be approached with a base of knowledge	Can lack a base knowledge or awareness
Are manageable; there is an element of control	Are unmanageable; there is lack of control

certain activity and the possible risks. The three questions posed above (What is wonderful about learning in nature? What do we need to be careful of? How will we stay safe and care for each other?) can be used for assessing risk in activities such as, for example, something as simple as sledding or playing chasing games on uneven surfaces to something more complex such as having a celebratory fire or using tools.

We assess and document activities that involve risk by using the Benefits of ECO form, which is shared below (for a blank form that can be photocopied, see appendix 2). These activities may include playing with sticks, climbing on a log or boulder, sitting around a fire, or being outdoors in inclement weather. This form is to help guide outdoor learning so the experiences are meaningful and safe for students and teachers. This template can be used for planning an outdoor session ahead of time, as well as for reflection at the end. This form can also be used to communicate the benefits of certain activities with teachers, school administrators, and the surrounding community. The following is an example of a new type of risky play being supported by introducing students to climbing on a large fallen tree in an outdoor setting.

Amy Butler

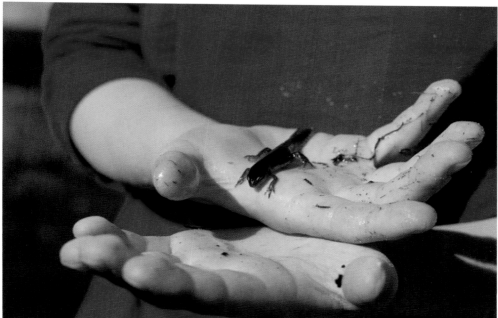

Emily Seiffert

being invited. In what ways do we get to know someone new? How are friendships made and sustained? We can apply these questions to developing a relationship to the natural world. In what ways can we get to know nature and care for it? You will find lessons throughout units 1 through 5 that incorporate acts of caretaking in the form of community science projects, learning about local animal species, and sharing gratitude. The lessons also help children to learn how to manage risk while building traits such as self-reliance and resiliency.

Emma Harris, the award-winning journalist and author, said in her TED Talk titled "Nature is Everywhere": "We have to let children touch nature, because that which is untouched is unloved." We can't expect students to fall in love and care for the Earth without touching it and, on some days, getting covered in it. When we love something, we care for it. The author and scientist Robin Wall Kimmerer reminds us that when we say, "I care for you" it also means I am going to *take care of you*. The evidence of the Earth caring for us is everywhere, and we can practice giving back and reciprocating this care in the very way we learn with nature. The three cares bring us back to remembering that caring for ourselves allows us to care for others. And when we are kind and caring for one another, we are in much better position to care for and protect the natural world.

Unit 1

Welcome to the Outdoors, Welcome to Nature

*One thing I've learned in the woods is that there is
no such thing as random, everything is steeped in meaning
and colored by relationships, one thing with another.*
—Robin Wall Kimmerer

The first weeks of school have passed. Where there once was boundless energy left over from summer vacation, the children are now settled in and moving through the daily rhythms of the school building. The schedule of specials, lunch, recess, and out-of-class services is established. It's almost as if the walls of the classroom expanded slightly with the students' returning energy, and then, in an instant, they quickly contracted back to order. Now, the school day ticks by—minute by minute, hour by hour.

Outdoors, summer is fading into autumn and warm days are still abundant. During a twenty-minute independent reading block, students gaze longingly out the window at a bright blue sky. There must be a breeze because the leaves are dancing. Yet no sound is heard as a single red leaf floats to the ground. Inside the classroom the air is stuffy and stale. A student is distracted and watches a little brown bird hop across the empty playground. The bird alights in a shrub near the baseball field. The student begins to hop in their seat while simultaneously staring at their book. Their arms raise up and down in a flapping motion.

The silence in the classroom is broken with a desperate plea.

"When can we go outdoors? Is this almost over?"

Another student speaks out, "Yeah! When can we go outside to do science and nature?"

This is the day students will exit the building after lunch for their weekly journey to learn outdoors, and their excitement is palpable. While the students look forward to a change of scenery and hands-on learning, the teachers understand the value of heading outdoors. In preparation for teaching beyond the walls of the classroom, the teachers have prepared by asking themselves, "What will we teach outdoors?" and "How will we know if the students are learning?"

These questions have brought us here, to unit 1: Welcome to the Outdoors, Welcome to Nature. The lessons presented here will get you and your students started on the path of learning in nature and with nature. These lessons help create an outdoor community where students will be engaged and active. Bundled together, these lessons support inquiry and exploration and help to build empathy in a wild space, and they also serve to help students and teachers get comfortable working together in the outdoors. This growing comfort and confidence is an important foundation for the lessons coming up in the following units of this book.

Lesson 1. Mouse Houses

Narrative by Amy Butler

At the base of a towering hemlock tree, a child is hunched over and focused on a very small task. Their hands move quickly, breaking larger sticks into smaller pieces, attempting to make each stick the same length. The precision of each crack is met with serious attention.

In deliberate fashion, the sticks are then propped up vertically on a tree root. The student pulls a bundle of milkweed fluff from their pocket and gingerly places it under the sticks. After a long pause, they stand up and dash toward a sunny spot in the forest. The child gathers golden maple leaves from the forest floor. With arms filled they explain to their fellow classmates, "You gotta go over there where the sun is for leaves!"

Upon hearing their classmate, children leave their projects and race toward the valuable resource. Chattering along the way, some children gather last summer's green ferns, others jump onto a favorite boulder, and one child stops and looks up into the canopy of the forest. "This is a dangerous place for a mouse! We need to be quick! Watch out for owls!" declares the kindergartener. There is a frenzy of collecting as the children scurry back to their projects.

One child shouts, "Mr. Parker, come see! Come see! I built a trap in case the fox comes, and this is where the mouse sleeps, and this is the kitchen for its seeds, and can it have a bathroom? Here is the back door. I want to make it a playground too. Can I get my journal so I can draw a picture of it?"

What are these children doing? They are building houses for mice.

Over the past thirteen years of working with public school teachers through Educating Children Outdoors (ECO), a program of the North Branch Nature Center, there has grown a collective understanding of the importance of allowing children to build during our time in the forest. This is an activity that never tires and fills the desire to play and imagine. Children are also very good at building. From the hunting and gathering of palm-sized materials to the beginning trials of an engineer, children will immerse themselves into any building project small or large. Children seem to be hardwired for building imaginary worlds they can put themselves into. There is no better proof of this than the explosion in popularity of fairy houses here in the United States. While the students creatively build, we as teachers feel pressured to meet performance expectations and standards. So, in an effort to meet the needs of both the teacher and the child, we created a lesson called "Mouse Houses."

To prepare and then fully integrate the experiences happening outdoors, teachers utilize their classroom time indoors as well. We have found that simply going into the forest without proper scaffolding can set students and teachers up for some level of frustration. To make "Mouse Houses" an integrated lesson, we aligned it with a classic kindergarten science standard. The teachers have begun their unit on living versus nonliving things. Before the children have an outing, the class discusses what they know about living and nonliving things. The children know that living things need food, water, air, and shelter.

We then ask them if these needs are the same for a mouse. The children have many ideas on this (and we help them dispel the common notion that all mice eat cheese), and the teachers then guide the students toward thinking about where mice sleep and raise their young. By reading and watching videos about different species of mice, the kinder-gartners notice the different materials used for building nests. During a short outing to walk around the schoolyard, children apply their new knowledge of mouse nests by making predictions about which natural items they find would be good for building a nest. Later, back in the classroom, students make a list of the mouse house materials they found. They also draw plans of what their mouse house might look like—just like an architect! Finally, construction day arrives, and the students bring their lists and plans into the forest with them.

The discovery of the needs of living things and the building of mouse houses does not end once the homes are built. Children come back during each outing to check on their homes, and each time we add on more opportunities for au-thentic learning. The children ask, "What if our mice get hungry?" We pass out handfuls of birdseed, and these are left behind in and around the many mouse houses. The following week children notice that the seed has disappeared! Where did it go? Who ate the seed? Of

Leah Greenberg

course, more inquiry develops around the missing seed and who the other possible inhabitants of the forest are.

Children asked if they could get some mice to put into their houses, so we introduced the use of clay to create life-sized mice. We found that pinecone bracts make excellent ears, and each mouse received a total of ten white pine whiskers. As we moved through each body part of the mouse, we learned more about how mice have adapted to survive and where in the forest to find the skinniest sticks to give jumping mice a proper tail. With their mice securely bedded down in their homes, we wave goodbye and thank the forest for our time playing and exploring.

Back in the classroom the teacher reads *Whitefoot the Wood Mouse*, by

Thorton Burgess, as the class transitions to the end of day for dismissal. The children are rosy cheeked and relaxed, lying on the carpet in the meeting space. They are captivated by this story and are undoubtedly invested visually and therefore imaginatively. One child has fallen asleep with their head rested on a backpack.

The kindergarten students are now caretakers of their mouse houses. Because of their continuous visits and observations, they have developed a strong relationship with a very special mouse and with a very special place in the forest behind their school.

Mouse Houses

Grade Level: K–Second

Objectives
- Students will learn how and why mice build nests.
- Students will explore a diversity of materials that can be used to make nests.
- Students will understand the needs of a living animal.

We Are Exploring These Academic Standards

NGSS.K-ESS3-1. Use a model to represent the relationship between the different needs of plants or animals (including humans) and the places they live.

NGSS.K-2-ETS-1-2. Develop a simple sketch, drawing, or physical model to illustrate how the shape of an object helps it function as needed to solve a given problem.

CCSS.Literacy.K-ESS2-2. Use a combination of drawing, dictating, and writing to compose informative/explanatory texts in which they name what they are writing about and supply some information about the topic.

What

Building small worlds is a timeless and favorite play schema of children no matter where they live (Sobel 2008). The drive of children to understand the bigger world around them can be processed through playing in miniature. We see this in the popularity of constructing fairy houses, playing with pocket-sized figurines, creating homes for stuffed animals, or through the digital world of Minecraft. In keeping to the learning expectations that need to be met in a public school setting, we can learn about the needs of living things by exploring mice through the lens of small worlds. Mice are not only universally known by all people, but they also offer much to study in adaptations, engineering, predator-prey relationships, and the food web. From the city mouse to the country mouse, what do these tiny creatures have

Amy Butler

to teach us from their own small and wild worlds?

How

Begin with asking the children what they already know about mice. The first answers you will get are that they eat cheese and live in people's homes. We live surrounded by the common house mouse and deer mouse. The number of different kinds of mice in the world is astounding, coming in at about 1,300 species! The understanding of fictional mice versus actual mice provides a place for connecting to the needs of a living creature.

Discuss what a mouse needs in order to survive. A helpful acronym to remember is FWARPS: Food, Water, Air, Reproduction, Protection, Shelter.

Ask the students, how do mice get their needs met in order to survive? Where do they live? How do they find and build a shelter? This is a good lesson to introduce in the fall as children can empathize with the work mice need to do to survive the winter. Start in the classroom with a variety of materials ready for children to explore what a mouse might use to build a nest. Leaves, grass, milkweed fluff, shredded up bark or plants. Include person-made materials as well, such as paper, cloth, string, and so on. Ask the students, why would mice use these specific materials? What is each material good for? Discuss insulation, waterproofing, and protection. Students can sort materials based on whether they are person-made or natural. If students decide to include people-made materials, be sure to practice a Leave No Trace ethic outdoors.

Students will be building mouse nests outdoors in an area that can be visited over time during the school year. A forested area, field, or protected area of the schoolyard will work.

They can begin by gathering materials they would like to use to build their mouse house. Explain how to gather materials ethically, by not taking too much and leaving little impact. Once students have their materials, they will find a place to build their house. Giving ample time for building allows children to really sink in and create what they would like for their mice. Do not be surprised when they start to build kitchens, garages, and bedrooms. This is a very important element of this lesson! It allows them the opportunity to create and organize their own world through constructing and building.

When children feel they are finished building, they can come and check with an adult and receive a small handful of sunflower seeds to leave in and around their mouse houses.

Students may also like to give a tour of their houses to other classmates.

Have students use their journals to record what their mouse house looks like and where it is located.

We Can Use These Materials Outside

Field guides for your region on mammals, pictures of different species of mice and mouse nests, natural and people-made materials for nest building, birdseed.

When We Get Back Inside

- Mouse houses are a great way to observe change over time. Each week students can take a short trip to visit their houses and record in a journal what has changed.

- Children can create a mouse to scale size with modeling compound or clay. Print photos of life-sized, or larger than life-sized mice so children can make detailed observations of animal adaptations before building.

  - What is the size of the eyes in relation to ears?
  - Are the back or front legs larger?
  - How long is the tail in relation to the body?
  - How long is the mouse's fur?
  - What is the nose length?
  - What is the color of their fur?

- Author Edna Miller wrote a series of books about a mouse named Mousekin. Our favorites are *Mousekin's Golden House*, *Mousekin's Family*, and *Mousekin's Lost Woodland*.

Ms. Burroughs' top ten tips for bringing nature names into the classroom

1. Work with the school librarian to gather many nonfiction texts on native animals.
2. Read nonfiction texts about each student's nature name.
3. Record animals you see when outside and add to the list throughout the year.
4. Watch live animal cameras of student's nature names.
5. Watch short teaching videos of student's nature names.
6. Create imaginative play opportunities involving nature names at play stations.
7. Play games involving nature names.
8. Include nature names attributes or characteristics in movement breaks.
9. Publish a whole class or individual student book(s) on nature names.
10. Invite families to gather for a celebration and reading of the class's published book(s).

similar but different. The search continued for the hare and the original plans for our morning in the forest were replaced by lessons with and from nature.

For many students, their nature name is their biggest connection to the outdoors and what inspires them to make the trek to their outdoor classroom week after week.

Throughout my years as an outdoor educator, I have seen many students with various disabilities eagerly participate in ECO and overcome their struggles. I have observed students with gross motor delays eagerly play in a game of mouse and fox because of the connection to their nature name. Students who have learning disabilities have completed the informational writing unit without complaint because they were invested in writing about their nature name. Children on behavior plans, time and time again, transform when invited into the outdoor arena. Nature names inspire students to act outside their comfort zone, to do hard things, and to persevere.

At the end of the school year, students are truly passionate about their nature names. By investing in these indigenous animals, children develop a longing to care for and protect these species as well as to share their knowledge with others. As academic as this research is, year after year, I am impressed by how studying their nature names lends itself to helping students progress in their social curriculum. I see diverse friendships form for the sole reason that two friends have chosen similar nature names. The changes in social interactions, for me, is one of the greatest benefits of this work

Hare, naturally took on the teacher role and called her friends over to share her findings and help educate her friends on what hares eat in the winter when there is not much food. Students ran off and started looking at tree branches to see if they had been nibbled. Another friend whose nature name last year in kindergarten was Eastern Cottontail enjoyed the joys of this journey, remembering that her nature name was very similar to Snowshoe Hare.

A whole class debate emerged around whether these two animals were the same or different. Eastern Cottontail herself clarified what indeed made these animals

Lesson 6. Sit Spot

Narrative by Amy Butler

"Expect the unexpected."

This is a saying I quickly adopted when I started learning with children while being immersed in nature. What I mean is, expect your plans to go sideways when a squirrel displays its acrobatic time outdoors and wish for it to happen more than once a week.

Spending time in nature also brings us into better connection with ourselves—children and educators alike—and that is yet another place

Nick Neddo

abilities above the students' heads during a focused lesson. Expect that a student will find a worm and then another and another and want to share these wiggling friends immediately with the entire class, right at the moment when you are giving directions. Know that students will lose a boot in a mud puddle and fill their backpacks with rocks and sticks to take home. Expect that children will love this to expect the unexpected. Expect that children will learn new emotional regulation skills. And then, unexpectedly, you will be surprised when your coworker or an adult volunteer tears up during the class's sharing time at closing circle. This reaction is a result of the soothing balm that gets generously applied when we spend time consistently in nature. These are the feelings that come to the surface when we sit quietly with ourselves and with children in nature.

A great way to develop this connection to nature and ourselves is through the use of sit spots. These are personal spots selected by each child and adult, where they can sit and observe nature or take part in a quiet activity. Over the years I have witnessed hundreds of children adopt sit spots and spend time sitting quietly in the forests, fields, and wild spaces in their communities. And during this time, countless adults have joined us to participate in this experiment of children immersing in nature during the school day. Once our routines for being outdoors as a class are honed, we invite parents, caregivers, and sometimes community members to join us. The result is a fantastic ratio of adults to children, and suddenly we have what resembles a village. It feels safe and joyous and still exudes moments of organized chaos.

The rhythm of our time outdoors is kept moving with the routines the students have embodied. Opening circle. Closing circle. Story and snack. Forest

Amy Butler

choice time and then the ribbon that ties it all together, sit spot. Sit spot is a core routine many nature-based teachers have adopted, inspired by Jon Young's work found in the book, *Coyote's Guide to Connecting with Nature*, and influenced by the millions of people who have sat in contemplation in nature for lessons and revelation. For these students, sit spot is the practice of sitting quietly (or as quiet as a child can be) at the same spot every time we have ECO.

When the newly initiated adults learn about sit spot, I can see the confusion and skepticism wash across their faces. What exactly are we doing? These kids are actually going to be quiet? How long do we need to sit for? Some parents reluctantly follow either their own child or one who has convinced them to join our sit spot routine. The small and tall human bodies fan out into the forest and sink into the landscape. It becomes increasingly still.

I hear a child shush a parent.

I see an adult lean back against a tree and close their eyes.

I watch a father hold his daughter's hand and squeeze.

We all can sense a peaceful wave roll over us. Slowly, each adult surrenders to the place they are in. A child digs a hole with a stick. The breeze picks up and I am positive I hear a collective sigh.

When the drum beats or a coyote call is given, children jump up and dash to the closing circle, eager to share what they saw at their sit spot.

The parent who got shushed tells their child how much they love them and thanks them for asking them to come today.

The parent leaning back against the tree covers their face with both hands and dries wet eyes.

The parent holding their child's hand does not let go, all through the closing circle, all the way back to the building.

What on the surface appears as a quaint activity to do in nature (or at worst a management tool) is secretly the Earth's most powerful healing modality. Sit here in a space with no ceiling and no artificial lights. Choose a spot to return to time and time again and breathe deep. Sit spots stay with the children through the rest of the day. The act of sitting in a safe and natural setting lowers cortisol levels and increases serotonin and dopamine. During sit spot, the forest is doing its work on the children and the adults, too.

Our time has come to a close and we are transitioning back into the structured school day. In our closing circle we share what we are grateful for from our time outdoors. Sharing gratitude threads together the day's activity and the attachment the students have to this place. I hear reflections: "I am thankful for playing with sticks," or "I am thankful my dad got to come today." Another child says, "I am thankful for my sit spot."

Our routine of sitting quietly every week outdoors connects us to the land and to each other. Then, practicing gratitude is the act of giving back. All of it serves to better prepare us for the rest of life. What would we have to give without connection?

Sit Spot

Grade Level: K–Third

Objectives

- Students will adopt their own place to visit each time they are outdoors.
- Students will make observations about the changing world around them.
- Students will be supported to record inquiries and observations.

We Are Exploring These Academic Standards

NGSS.K-ESS-2-1. Use and share observations of local weather conditions to describe patterns over time.

NGSS.1-ESS1-2. Make observations at different times of year to relate the amount of daylight to the time of year.

NGSS.2-ESS1-1. Use information from several sources to provide evidence that Earth events can occur quickly or slowly.

NGSS.3-LS4-4. Make a claim about the merit of a solution to a problem caused when the environment changes and the types of plants and animals that live there may change.

What

What is a sit spot? Your students may already have areas they are drawn to in the outdoor learning spaces you visit frequently. A boulder, a fallen tree, or a spot where the sun shines through the canopy. These favorite places to play and engage become landmarks in a wild space, and this familiarity of place is a good foundation for choosing a sit spot. A sit spot is a special place that is adopted individually by each student (Young et al. 2010). This is a place in your outdoor learning area where students visit at the end of your planned time outdoors. Sitting quietly and independently at the end of time together allows students to quiet their minds and bodies in preparation for reentering the school building. Teachers need it too! In essence, time at

Inspiring and supporting sit spot

- Draw a sit spot plan. Where will you sit? What would you like it to look like?
- Adopt a tree bud at your sit spot and document the changes each week.
- Fox-walk or sneak up on your sit spot. Walk or crawl a different way to get there!
- Engage all the senses. Count sounds. Sit blindfolded. Bring tea to drink. Make a list of textures and smells.
- Find and label the cardinal directions. Where does the sun come up?
- Hang a feather from a string and watch the direction of the wind.
- Make a map of where you are sitting. Where are other classmates sitting?
- Children can pretend to be sleeping fawns until the doe (the teacher) comes and wakes them.
- Write a thank you note to your sit spot.
- Draw a picture of your sit spot at night.
- Invite a friend to join you. Invite a family member.

our sit spots is an opportunity to practice mindfulness and assimilate all that nature has offered us. The active buzz and activity of the learning and exploration slows to a hush. A sit spot affords children and teachers time to self-regulate by co-regulating with the breeze, the trees, bird song, and a sense of knowing this quiet space will be here next time.

How

Finding a sit spot does not require a tract of pristine forest. Sit spots can be in the schoolyard, under a grove of trees in a park, in a hedgerow between neighborhoods, or even along the edge of an athletic field. These are the nooks and crannies of the natural world where we can find birds, observe the weather change, and watch plants grow. Once you have decided on an area that is easily accessible from your classroom, guide students to choose a place where they can sit by themselves. The key word here is "guide."

Students are encouraged to find a place that is comfortable to them and close to something they find intriguing. This may take some time. It is okay for it to take a few days to find just the right sit spot. Much as in *Goldilocks and the Three Bears*, some investigation pays off! The act of choosing your own sit spot can be ceremonial or not. We can set students up for success by scaffolding the experience with stories, games, and provocations that inspire them to become friends with this special place they have chosen to visit every day.

It is important that adults can see every student from their own centralized sit spot. Younger children will naturally want to be closer to each other and older students farther away. Find a happy and safe medium that will work for you and your students to fully gain the benefits from a quiet sit spot time. Preschool and kindergarten students might practice sitting quietly as a whole class. If you practice mindfulness in the classroom with students, use this as scaffolding for sit spot time. Introducing games that aim to teach stillness and stealth are great ways to build skills that can be applied when sitting quietly and independently.

Sit spots are part of the ECO schedule at the end of our time outdoors. Children are warm from playing, satisfied with snack, and have had a full session of learning and exploring with their peers. Do not be surprised when children ask if it is sit spot time yet! That is a clue they are ready to rest and connect with their own quiet place. Call students back to a closing circle to end your time outdoors and invite them to share stories from their sit spots. Write down what they share, because this is the story of place and of their relationship with it.

We Can Use These Materials Outside

Students can go empty handed to their sit spots, or they can bring something to help them sink in and engage. Journals are a great option for busy minds. Keep journals in resealable plastic bags and be sure to include pencils and colored pencils as writing utensils. Pens will freeze and do not work well when wet. See the above text box for reminders on what you could bring.

When We Get Back Inside

Observations made at sit spots can be tracked and further explored back in the classroom. An important consideration here is to let sit spot be what it needs to be for your students and to not turn the time into an assignment. If there is genuine interest in observations made during sit spot, support students in researching and learning more.

Other students benefit from this routine to stimulate the parasympathetic reaction that helps to calm us down. As teachers we can prioritize what our students are gaining in health and wellness over meeting expectations of learning during sit spot time. The feeling of being "alone" in nature is very different from sitting quietly at your desk inside a building. Sit spot outdoors cannot be replicated indoors. Take a moment to think, how often are you invited to sit quietly outdoors and breathe deep during a school week? Would you like to be invited? Your sit spot is calling you!

Books about Sit Spot

Me and My Sit Spot, by Lauren MacLean

Coyote Creek Alliance—Milo's Sit Spot, by Maryam Moss

References

Sobel, David. 2008. *Childhood and Nature: Design Principles for Educators.* Portland, ME: Stenhouse Publishers.

Young, Jon, Ellen Haas, and Evan McGown. 2010. *Coyote's Guide to Connecting with Nature.* 2nd ed. Santa Cruz, CA: OWLink Media.

Unit 2

Build It and They Will Come
The Power of Sticks

Sticks! Sticks! Glorious sticks!
They are perfect for building,
they're thin and they're thick.
When they're as long as your arm, they do harm,
So, take lots of care when you're building with sticks!
—*Mary Zentara*

"Do we let them play with sticks?" This is one of the very first questions before going outdoors with a classroom of children. We wonder; what does it look, feel, and sound like if twenty children are wielding sticks in the forest? Unit 2 is a collection of lessons that break down the barrier of "no playing with sticks" and a place where we can begin to observe the benefits of allowing children to construct their world physically, socially, and emotionally through stick play.

These lessons help to scaffold stick play safely while meeting teachers' comfort level in exploring how sticks can be used as a tool for learning. What might we expect the outcomes to be of all this stick play? Here are just a few: building positive interaction and communication skills; learning problem-solving strategies; developing spatial awareness; strengthening physical skills; appreciating what your body can do; and, for some children, it gives them an opportunity to develop self-awareness and improve concentration by moving, manipulating, and orchestrating objects they have found in nature. So, what if a child picks up a stick in the forest? We might be more concerned if they didn't!

Lesson 7. The Golden Stick: Measuring Sticks and Stick Safety

Narrative by Harriet Hart

In November, the forest floor of our outdoor classroom transforms and becomes full of loose parts to enjoy. Newly fallen leaves of all shapes and sizes are in abundance. Sticks and rocks become exposed as leaves are piled by busy hands. There are many opportunities for play and exploration. This morning's preschool class was keenly experimenting with all that nature had to offer.

In the spirit of investigation, one of the students picked up a rather large stick. This stick quickly became a tool for dislodging rocks in a mound of soil. And I noticed the stick was also a risk to the surrounding students, being that it was long, cumbersome, and not within this student's full control. We talked about how to safely manipulate such a large stick. We modeled how to safely use it. But, as playtime drew to a close, we realized that we were going to need a more inspiring method of supporting safe stick play with this class of young learners.

In our planning time the following Monday, our teaching team talked about how our students seemed drawn to stick play, what made us nervous about it, and the possibility of welcoming it into our classroom. It was easy to imagine the accidents that might happen when playing with sticks, but we found that the benefits—including opportunities for problem-solving, engineering, and imaginative play—outweighed those risks and the likelihood of a serious injury.

Sticks are an open-ended natural material that we felt were a valuable loose part; we wanted to find a way to say yes to our students' interest in them, but we felt that we needed a scaffolded approach. We considered how to best support our students' developing spatial awareness and examined our own comfort level with stick play and its risks. We decided to support the interest in stick play by having students interact with shorter sticks first. We found a stick length that we felt would be manageable for our students, one that would reduce some of the risk while preserving the benefits, a stick that would allow for maximum independence and minimal risk. To give our new boundary some appeal—some gravitas—we painted this stick gold.

The following week we introduced it to our class quite simply as a new rule: "Any stick that you want to use in the forest today needs to be shorter than the Golden Stick. The Golden Stick will stay here at the stump circle so that everybody can see it."

For the next hour we watched as our students took our new rule and turned it into an incredible math lesson. We had expected to have to remind students about our expectations and prompt them to refer back to the stick we had chosen. Instead, we stood back as they spent the entire choice time finding sticks and bringing them to the Golden Stick to measure and compare. They lined the sticks up, noticing whether the ends stuck out further than those of the Golden Stick. As they hunted for more sticks to measure, they traced the pathways of the outdoor classroom, mapping the space with their movements. They looked closely, experiencing the varied browns of the forest floor, the scent of pine needles, the stickiness of sap, and the sound of wind in the trees. The process of measuring captivated them, demanding their attention and motivating their learning. They were invested in their project, simple though it appeared. The gold paint had successfully inspired them.

Since that November, the Golden Stick has continued to evolve as a tool in my teaching. When the timing seems right or necessary, I arrive at circle time with a mystery bag and a story on the tip of my tongue. My students are drawn into the tale of a little beaver who is anxious to learn how to build but cannot find sticks of just the right size and weight to be successful. I open my bag and pull out the Golden Stick; the story becomes real, and the little beaver is presented with a solution to the problem (see "The Story of the Little Beaver and the Golden Stick" in lesson 7).

Using an oral story to introduce our stick safety rules addresses foundational literacy and promotes listening skills. Stick safety is the forefront of the lesson, and the Golden Stick is a starting place for those who need more structure and guidance on how to safely manipulate these materials. And of course, math continues to be central to the unfolding. It is the place where math and magic intersect.

Weeks later, our formal stick lesson behind us, we were walking through the woods. We came to a place where the path curved around a large boulder,

Amy Butler

upon which a tree was growing. Its roots wrapped the contours of the rock like a claw before disappearing into the ground. Its bark was papery and curling, yet radiant. The tree trunk was golden. As the class explored the nooks around the base of the boulder, one of my students approached me. He had a significant speech delay that made it especially challenging for him to express himself and much of the time he was hesitant to speak, but today he urgently needed to tell me something. Getting down on his level I listened and prompted, and I finally realized that he was bursting with the excitement of trying to tell me that this tree must be where our Golden Stick came from! The magic of the Golden Stick was suddenly evident. This special stick was important to our class community; it had become an essential part of our narrative and it was more than just a rule. When we thought we were simply scaffolding a needed safety protocol, we were, in fact, introducing math and magic all in the context of stick play.

The Golden Stick

Grade Level: K–Second

Objectives

- Students analyze and interpret the length of an object.
- Students will plan and carry out an investigation involving measurement.
- Students will use addition in multiples of five and ten to strategically gain points.

We Are Exploring These Academic Standards

CCSS.Math.Content.K.MD.A.1. Describe measurable attributes of objects, such as length or weight. Describe several measurable attributes of a single object.

CCSS.Math.Content.K.MD.A.2. Directly compare two objects with a measurable attribute in common to see which object has "more of" or "less of" the attribute, and describe the difference.

CCSS.Math.Content.1.NBT.C.6. Subtract multiples of ten in the range between ten and ninety from multiples of ten in the range between ten and ninety (positive or zero differences), using concrete models or drawings and strategies based on place value, properties of operations, and/or the relationship between addition and subtraction; relate the strategy to a written method and explain the reasoning used.

CCSS.Math.Content.2.MD.B.5. Use addition and subtraction within 100 to solve word problems involving lengths that are given in the same units, for example, by using drawings (such as drawings of rulers) and equations with a symbol for the unknown number to represent the problem.

CCSS.Math.Content.2.MD.A.1. Measure the length of an object by selecting and using appropriate tools such as rulers, yardsticks, meter sticks, and measuring tapes.

What

The story of the Golden Stick is a brilliant example of how to successfully support the most common form of risky play in an outdoor setting. By using the power of storytelling to captivate and motivate children, we give them the opportunity to safely explore sticks and all their many uses and applications. Not only does this story and corresponding lesson build a foundation of confidence and self-regulation within the act of stick play, but it also models mathematical reasoning and asks students to think quantitatively and strategically. Sticks are an open-ended loose part that have endless creative possibilities for all students.

How

You will need a very special stick that is painted gold for this lesson! The stick should be the length that you are comfortable with your students first using. A Golden Stick more than one-half inch in diameter will ensure that it will not break and will be able to handle the attention it will receive. A preferred stick for painting is one that is dry and has most of its bark removed. You might consider wrapping golden-yellow yarn or material around the stick for added flair.

Kelsey LaPerle

You can make Golden Sticks of various lengths to prompt a math lesson and extend what has been taught inside the classroom to the outdoors. For our youngest learners, simply comparing sticks they have found to the Golden Stick is the focus of the lesson. We can take this time to observe how students navigate the terrain looking for sticks, how they gather and move with sticks safely, and, in the end, how they choose to use their sticks. For older students, a math game can be created with a point system based on the Golden Stick that challenges students to get to a final score of one hundred points or more. For example, finding a stick that is longer than the Golden Stick earns ten points each time, a stick that is the same length earns twenty points, and a stick that is shorter earns five points. Create your own point system and add other challenges that require students to use subtraction. Use the Golden Stick to prompt building two-dimensional and three-dimensional shapes and to identify their attributes. Sticks are a renewable and available math manipulative that can be used safely by every child.

We Can Use These Materials Outside

- If you do not have access to sticks, they can be imported to your location. Whether it is an open field, a fenced play area, or an asphalt yard, sticks can be used in a variety of non-forested outdoor settings.
- Consider using rulers, tape measures, yardsticks, and metersticks for determining the actual lengths of sticks.
- Bring string for tying sticks together to make three-dimensional shapes.

The Story of the Little Beaver and the Golden Stick

by Harriet Hart

It was spring and a beaver family was exploring their pond after winter. Little Beaver watched Momma Beaver gnawing at trees and dragging big sticks to build up the dam and lodge. But when he tried to help, he found that he couldn't drag the sticks, they were too long, they were too fat, they were too heavy. Momma Beaver saw Little Beaver trying with all his might, and said, "One day you'll be big enough to carry sticks around, but not yet, Little Beaver."

Little Beaver was disappointed and a little sad; he hoped that he would grow fast so that he could help out. The spring sunshine felt warm, so he swam through the pond, thinking and dreaming, swimming, thinking, and dreaming.

As he swam, he saw something sparkling out of the corner of his eye. He turned his head to see, but it was just the sun sparkling on the water. He carried on swimming and thinking, and then he saw the sparkling again. He turned his head quickly to see what was there and noticed a sparkling that was moving away from him across the pond. He swam after it. It danced from wave to wave and from ripple to ripple, moving away across the water. Little Beaver swam fast, with his eyes open, chasing the sparkle. Suddenly it disappeared, and Little Beaver found himself looking up at a golden tree on the bank.

Little Beaver swam closer. The tree was gleaming in the sunshine; it was almost too bright to look at! One branch of the tree was hanging down close to the water; it was at just the right height for Little Beaver to gnaw at. Little Beaver used his large front teeth to gnaw away at the wood until, splash! It fell into the water and floated by his nose. Little Beaver picked up the stick with his mouth—it was just the right weight, just the right length, and just the right width for him to carry! For the rest of the spring, Little Beaver carried the golden stick with him, so that he could make sure to find the sticks that were just right and bring them back to help his family build up their lodge and dam.

- A plain piece of fabric acts as a clear visual background to lay sticks for demonstrating measurement and constructing shapes.

When We Get Back Inside

- Sticks that are the size of your Golden Stick can be brought back into the classroom and used in the block area or as math manipulatives.
- Linking your lesson to natural history offers more connections to the natural world through the exploration of tree species, studying deciduous and coniferous trees, and learning how native animals depend on trees for shelter and food.
- Start an ongoing list of things that children can do and have done with sticks in your outdoor setting. Write a classroom book about it!

Books about Sticks

Not a Stick, by Antoinette Porter

A Leaf, A Stick, and A Stone, by Maggie Felsch

The Stick Book: Loads of things you can make or do with a stick, by Fiona Danks and Jo Schofield

Lesson 8. Glorious Sticks: Scaffolding for Sculptures and Structures

Narrative by Jillian Zeilenga

The forest is aglow with crimson and amber foliage, and it reflects the afternoon sun through the tall birch and maple trees. It has been five weeks since school began, and the students have settled into our routines of learning in the forest behind the school.

After entering the forest, the children disperse for some self-guided learning, continues to put it in a special place when we go back inside the school building. They come back to it each day like it is their best friend. I have never seen a child take that much care with a pencil or a pair of scissors. Sticks are special. As one student put it, "You can do whatever you want with them! They are so cool!"

Our class has been busy with sticks

Nick Neddo

and I take to my "teacher perch" on my favorite fallen log to observe my students. From here I can see all the children within our outdoor classroom boundary. I soon notice a familiar and very popular object clutched in their little fists: a stick.

If there is one natural item that kids gravitate toward more than anything in the outdoor classroom, it is sticks. The excitement you see on a child's face when they have found their perfect stick is priceless. I have a student who found a stick on the very first day of school and since school began, and I saw this as a perfect opportunity to use sticks to introduce mathematical thinking through measurement and balance. A favorite forest activity of all children is building with sticks, and by introducing a stick sculpture challenge we can collaborate and work on problem-solving. This was the third lesson we had done with sticks, and the students were already aware of all the safety protocols and had been practicing our stick safety rules.

When the students gathered at the

Our stick rules

- When you are moving a stick bigger than yourself, say, "Safety circle!"
- When you carry a stick remember, "Low and slow."
- Use only sticks that are dead and on the ground.
- Sticks do not touch people or other living things.
- Sticks are not weapons. Sticks can be [fill in the blank].

Take care of your stick!

circle for our stick sculpture lesson, I asked them if they wanted to see my favorite stick. They all responded with an eager "Yes!" From behind my back, I carefully and ceremoniously presented my favorite stick. I explained to them that it was my favorite because it was smooth, it was the perfect length for me, and it had a cool bump at the end. I told them that they would be looking for their perfect stick today. The only rule was that it could not be taller than themselves. The students scurried off in search of the perfect stick. As the students found various sticks, they were holding them up to themselves to measure the length and height.

"Is this one taller than me?" one student asked.

I could see other students breaking off sticks until they felt they were the right length. Some students carefully scrutinized sticks until they found the one that met their qualifications. As soon as a child made a final decision about a stick—"Yup! This is the one!"—it was easily changed upon seeing the potential of another stick on the forest floor. Sticks were being adopted and abandoned quickly. The students' ideas changed about what was a perfect stick, and the chatter among students was steadily increasing and their language getting more complex with each decision that was made. Our time hunting for sticks was coming to an end, and once they all had a stick, they returned to the circle and shared why they had chosen that stick.

After our share, I told them we would now be making a unique sculpture with all our sticks. It started with one stick. I dug my stick into the ground so it stood upright and then asked who would like to add their stick. Naturally all their hands shot up. One by one the students came and carefully thought where they wanted their stick to be placed on the sculpture. As more sticks got added, their thought process became more elaborate.

The students were problem-solving how to attach the sticks and soon suggested we use tape or pipe cleaners we had brought. The sticks were all different lengths and weights, so it required some planning on their part on how to execute their ideas.

"I want to put this one on top!"

"This stick has too much heaviness. It won't balance."

"We need a stronger stick up here."

As our sculpture grew and became off-kilter, there was a danger of it toppling over.

One student piped up and said, "We need a stick to stabilize it!"

There was a mass exodus from our circle and children quickly gathered more sticks from the forest. They tried sticks of different sizes to find one that would fit. A collective cheer filled the air as balance was achieved. Once the last stick was placed, we all took a moment to admire our work.

Making a stick sculpture together as a group provides a shared experience for the class that we can refer back to. It provides some scaffolding on how to get started and what next steps they could take. It gave students a chance to build upon each other's ideas and provide feedback.

After completing our class sculpture, the students went off to plan and make their own stick creations. Some students grabbed clipboards, paper, and pencil and sketched a plan. Others got to work right away hauling sticks and declaring, "Safety circle!" as their construction process started.

As children were busy working on new stick plans, two of the students spent their time adding to the original stick sculpture instead of creating their own. One made a plan on paper but then added to the existing sculpture. The other child was not yet comfortable making their own plan but had plenty of ideas on how to add to our class sculpture. This child attended the longest to their work. After watching the child's process and reflecting, I realized how important it was to have our group sculpture for him. This gave him access to the lesson and the independence to do some work on his own. He focused with deep attention on each stick as he added to the class

Amy Butler

sculpture. A fall breeze moved through the forest and the students settled into their play. There was the magical hum vibrating throughout the forest of children engaged in authentic and purposeful learning. I checked my watch—3:15! Time had slipped away once again in the outdoor classroom. As the children ran down the pine needle–covered path toward the school building, one child whispered to me, "I can't wait to come back tomorrow!"

Glorious Sticks: Scaffolding for Sculptures and Structures

Grade Level: K–Third

Objectives

- Children will understand and apply safety when using sticks for learning.
- Children will learn how to measure the length of an object.
- Children will understand how length, width, and weight affect how we can build with sticks.
- Children will work collaboratively as a group to create a stick sculpture.

We Are Exploring These Academic Standards

CCSS.Math.K.MD.A.1. Describe measurable attributes of objects, such as length or weight. Describe several measurable attributes of a single object. Measure lengths indirectly and by iterating length units.

CCSS.Math.1.MD.A.1. Order three objects by length; compare the lengths of two objects indirectly by using a third object.

CCSS.Math.2.MD.A.1. Measure the length of an object by selecting and using appropriate tools such as rulers, yardsticks, meter sticks, and measuring tapes.

CCSS.ELA-Literacy.CCRA.SL.1. Prepare for and participate effectively in a range of conversations and collaborations with diverse partners, building on others' ideas and expressing their own clearly and persuasively.

What

Scaffolding any type of risky play is our responsibility as teachers in an outdoor setting. This lesson is meant to do just that and guide children toward safe play with sticks larger than the Golden Stick (see previous lesson). By scaffolding larger stick play with a teacher-directed lesson, we can also tap into the agency that children feel when they find that perfect stick. Building a stick sculpture requires that we provide space and time to build with larger materials. With more space, it allows for more movement, which creates more language use and, as a result, deeper cognitive development.

Before starting this lesson, conduct a site assessment of the different types of sticks you have access to in your outdoor classroom space. Are there enough for children to use freely? Are they easily accessible? If you are utilizing a public space such as a park, consider asking permission to utilize sticks found in that space. We have made signs in the past labeling our work so other community members can learn about the students' experiments. This goes a long way toward connecting the greater community back to the school.

If you do not have immediate access to sticks in your outdoor learning space, bring them in! Ask within the school community if someone can donate a pile of sticks. Asking for stick donations sounds odd, right? But you will get endless learning opportunities with a sizable pile of precut and delivered sticks. A final resort to stick acquisition would be to seek out a lumber yard, landscaping company, arborist, or florist as a source.

How

First, start a discussion as a class about safety when gathering and moving sticks. A helpful way to scaffold stick exploration is to ask these three questions and write down children's answers:

> What is wonderful about sticks?
>
> What do we need to be careful of when using sticks?
>
> How will we stay safe and keep each other safe when we are playing with sticks?

Sticks that are the size of your hand and forearm can be brought into the classroom for exploration. Children can make letters or numbers, spell their names, or build mini structures with sticks inside the classroom before you head outdoors.

When children are going to use sticks bigger than themselves, have students practice making a safety circle around themselves by saying out loud, "Safety circle!" This lets classmates know that you are about to move with a stick that is longer than your body.

When safety rules are set and agreed upon by the class, split the class into small groups and have them go into the forest to find a stick that is no taller than an adult in your group. The students will bring their sticks back to a designated meeting spot in an open area.

Once everyone in the class has gathered back at the meeting spot with

Amy Butler

their sticks (have everyone sitting with their stick at their side), ask: Who thinks they have a short stick? Who has a long one? Who has a medium-sized stick? How can we tell?

Next, you can do several different things, depending on the age of your students. Estimates could be made: How many hands tall? How many fingers tall? Have tools for measurement and the children can measure their sticks. Once they have, they can report to an adult who will write it down on a white board for everyone to see. This can then lead into a discussion of who has the longest stick. This could also be an opportunity to add and subtract lengths. You could have the children put their sticks in order from longest to shortest. This is a bit of a movement and management challenge. Have children estimate visually the sticks that are longest or shortest. This will help in not having all the children getting up at once with their sticks! You could also have your children sort their sticks depending on different attributes, such as length, color, or width. Have your students decide how to organize them and how to do it safely.

After the longest to shortest sticks have been determined, use the three longest sticks to make a tripod. This is a great discussion on what a tripod is and how it works. Have the children hold their sticks while an adult lashes them together with string or rope. This tripod will be the base for the stick sculpture. After the tripod is stabilized, each student may come and place their stick on the tripod. After all the sticks are placed, step back and admire your sculpture!

After you introduce this lesson, the students will want to build. This lesson is intended to be used as a scaffolding activity for future shelter building activities. The suggestion is to have a bigger building project planned for another session—expect this to happen. Once sticks are introduced, there is no turning back!

We Can Use These Materials Outside

- String can be an option after some initial attempts at sculpture design have proved challenging. Offering students a length of string or fuzzy wire for securing sculptures and structures adds to the possibilities for successful engineering!
- Set up a string cutting station and have scissors on hand so children can cut their own string and use it to add to their creations. Discuss with children how string can be used in the setting and expectations for cleanup.
- Have pictures of inspiring ephemeral art such as that by Andy Goldsworthy to inspire stick creations.

When We Get Back Inside

- Since sticks will now be used often in your outdoor setting, review with your students how can sticks be used safely. Make a Safe with Sticks rules poster and laminate it so it can be brought outdoors as a reminder.
- For future outings have students make a stick plan. Students can draw and label plans that can then be put into a plastic sheath and brought to the outdoor classroom.
- Students can write a how-to for building a stick sculpture for another class. After sharing stick building instructions, have students come join the experienced stick engineers outdoors for exploration.
- Photo-document stick play and display photos in the classroom. Photo documentation provides visual scaffolding for students as well as providing them with a narrative to start for those who may have difficulty expressing themselves verbally.

Books about Sticks

Messy Maths, by Juliet Robertson

Stick Man, by Julia Donaldson

Steve the Stick, by Noreen Greiman https://www.entangledharmony.com /free-story-a-stick-and-a-new-friendship/

The Organic Artist for Kids, by Nick Neddo—the "how-to" on turning sticks into paintbrushes, pens, and charcoal

Lesson 9. Cardboard Kids

Narrative by Amy Butler

We are well into spring in northern Vermont, yet the temperature outdoors reads 40 degrees at 8:30 a.m., and a forecast of rain is imminent. Just two weeks ago children ran about in the forest with short sleeves and asked to shed their mud pants. Spring had arrived

can get dressed for the weather, pack our bags, and hike all the way to our destination. These are the first steps to developing self-efficacy. With strong self-efficacy, students believe in themselves and are able to tackle tasks in their lives that are challenging. By being outdoors in all

Nick Neddo

and just as quickly seemed to be hiding in the shadow of winter.

The importance of creating opportunities where children need to care for themselves is foundational in our learning during ECO. What may seem like simple and everyday skills of dressing are emphasized in the self-care that needs to happen when students are learning outdoors. Getting those mittens on and zippers zipped is sometimes not easy. The social modeling of watching others master these routines helps children to believe that they can do the same. We

types of weather, we are demonstrating to the children that yes, we can do this, and this is a part of school. And we can do this together!

In a first-grade classroom, twenty-one children are gathered in a circle for a morning meeting. Small bodies jostle for space, and the teacher begins to sing a gentle song that brings the students' attention to the start of their day. The topic of the morning meeting is, "How can we take care of ourselves in the forest today?" After a few quiet moments, the teacher invites children to share with

Leah Greenberg

"I think we need mud pants or maybe we have to wear snow pants? Is it cold out?" And then, "There's no snow, so we don't have to wear snow pants."

"ECO socks! Don't forget ECO socks!"

These are all comments based on experience from a class of first graders who had been learning in the outdoors since kindergarten. Yet, spring in the Northeast is tricky, and we sometimes find our adult selves digging a little deeper to tolerate the fluctuations of the season. The children, on the other hand, are eager to get out the door and explore, regardless of the weather forecast. As a class, the decisions are made on what to wear for the day. Mud pants, hats, and definitely our coats.

Still gathered in our morning meeting, the bodies start to squirm. The teacher presents to the children a cardboard cutout of a small person. It stands about eighteen inches high and looks much like a gingerbread person.

"These are some cardboard kids that have never been out in the forest. They need your help to get dressed for being outdoors. How can we dress them for today?"

The students begin to work in pairs to adopt a cardboard kid and dress them for the day. Time is spent in the classroom coloring, cutting, and pasting. Soon the cardboard kids come alive with purple rain pants and yarn hair. As the students characterize the cardboard cutouts with faces, they are given names. A few students draw backpacks on their kids just in case they need a place to store their mittens. Soon it is time to head outdoors, and children are chattering away about

their neighbors how they will take care of themselves today in the forest. After they discuss ideas with their neighbors, the teacher asks students to share how they will care for themselves on such a chilly and wet spring day.

"We need snack and water because it's hard work out there."

"I have boots today. Do we need boots?"

"We have to be able to see an adult!"

"I forgot my hat and mittens because my mom packed them away because it's supposed to be spring. But maybe I need them today."

After singing the song, students wonder, "What's a thicket?" and "Hares have so many predators!" Another student asks, "How do they survive?" We pause to answer questions and encourage their inquiry, continually bringing them back to what we had learned earlier in the week inside the school building. These connections are so important for growing their knowledge and understanding of the world around them, snowshoe hares being a part of that world in their forested outdoor classroom.

After the song and discussion, the big announcement for the day is made. The children will be building slash piles of sticks and logs. Dave announces, "Hares need a safe place to hide from predators, and you can help."

The group is led in a guided discovery to determine how big and how wide the slash pile should be for a snowshoe hare to hide from its predators. Students finally conclude that it needs to be twice as long as an eagle's wings (between ten and fifteen feet) and at least as tall as the students themselves (from four to five feet)!

After accepting the challenge, eager to help the hares, the groups head off into the forest to find the right spot. Children can be heard saying, "Remember we learned that it should be near the pond. Let's head over here." And, "It also needs to be in a flat space near lots of sticks." Groups move off, ready to begin their job of helping snowshoe hares survive this winter.

One group heads to a flat spot with lots of fallen trees and sticks; perfect for construction! Other groups head off over the hill to a more secluded spot,

Leah Greenberg

"out of bounds" from our base camp. Children get to work dragging, hauling, and yanking large branches and dead trees over to a chosen spot. This takes a lot of perseverance and determination. It is not always easy to move logs twice as big as themselves. They do not give up! This dragging and hauling engages problem-solving strategies and builds core body strength in these first- and second-grade children.

We layer the branches and sticks, starting with larger logs at the bottom. Our Rabbitat is taking shape! Some children take charge, telling others, "Get five sticks as long as your arm," and "Lay that branch the other way!"

As the structure builds, we introduce two specific predators to our group that we need to consider while constructing the brush pile. How can we keep the hares safe from a hawk or an owl's talons? What could we put in the openings to keep the coyote out? The children examine their construction again, looking carefully for places these predators might be able to penetrate the fortress they have created. With determination, they fill in the gaps and add layers to what they have built. Working as a team, with the goal of protecting the hares, the group works with perseverance and care. They listen to one another's ideas in order to create the safest haven possible. These moments of teamwork and collaboration carry back to the classroom, where students work on tasks that feel very different. We can anchor their teamwork on these reflections from the forest, reminding the students of these positive encounters and experiences.

As they continue to work on building and fortifying the structure, some other students take on the role of decorators, finding birch bark to lay on their structure and beautiful leaves for a bit of color. They add other natural materials they find in the forest, wanting the hares to have not only a safe place but also a beautiful place to hide. Each child has a role in this task. Every ECO Challenge we engage in has opportunities for different learning styles and preferences to shine. This is nature-based learning. Everyone can find a way to contribute, be heard, and in the end be successful.

The weeks that follow are filled with excitement to see if their hard work has paid off. The students hope for snow so they can see if the Rabbitats have been put to use by the hares. Will we find hare tracks in and around our Rabbitats? Will we find predator tracks as well? Who else might be using these brush piles as a protective shelter on a cold winter night?

Rabbitats—Shelters for Furry and Feathered Friends

Grade Level: K–Third

Objectives

- Students will consider what animals need to survive, as well as the relationship between their needs and where they live.
- Students will determine if a habitat has adequate coverage for prey species.
- Students will design and construct a brush pile for prey species.
- Students will observe and document signs of use over the course of a season.

We Are Exploring These Academic Standards

NGSS.K-2-ETS1-1. Ask questions, make observations, and gather information about a situation people want to change to define a simple problem that can be solved through the development of a new or improved object or tool.

NGSS.K-2-ETS1-2. Develop a simple sketch, drawing, or physical model to illustrate how the shape of an object helps it function as needed to solve a given problem.

NGSS.3-LS4-4. Make a claim about the merit of a solution to a problem caused when the environment changes and the types of plants and animals that live there may change.

What

Creating brush piles to enhance habitat for prey species is an active way to get students engaged in learning more about a specific species. It can also inspire students to be caring stewards of the environment. Brush piles can be constructed in a school yard as shelter for overwintering birds or in a public park as shelter for turkeys, cottontail rabbits, or raccoons. Understanding the need for brush piles in a selected habitat is important to its successful use as a shelter. Be sure to inquire with your local park if a "rabbitat" would be needed and work with park officials to partner on the project of building one. Learning about the needs of species in your region builds upon children's innate desire to

Amy Butler

care for animals. This helps students to understand how an animal like the snowshoe hare survives, and ultimately it gives students a deeper understanding of the interconnectedness of themselves and another species.

How

Begin by having a discussion with students about wildlife survival needs in winter. What do they need? Where do different animals take shelter? How about small ground-dwelling prey animals, like rabbits, snowshoe hares, and grouse? What might we be able to do to help them survive?

Demonstrate building a miniature model of a rabbitat in the classroom. Using small props representing predators and prey, show how a slash pile functions. Use sticks of varying thicknesses to build a model on a table or desk to demonstrate its many entrances and exits and hiding spaces. With extra materials on hand students can build their own model and practice the sequencing.

Once you have received permission about where you can build a Rabbitat, spend some time surveying the area for signs of animals. Is this a good place for a rabbitat? Are there enough materials for building one? What about two? Split students into groups and have them determine where they build their Rabbitat. After practicing stick safety in previous lessons, this is a lesson that encourages bigger motor movements, motor planning, and collaboration. It also has a direct purpose, to protect snowshoe hares from predators. Students are invested and connected immediately!

We Can Use These Materials Outside

- Field guides about habitat, food sources, and signs of snowshoe hare
- Pictures of the sequenced building method for rabbitats
- Life-size cutouts of snowshoe hares to use as a reference for sizing entries and exits to the rabbitats. Cutouts can also be used to hide in the slash piles to determine how well animals may be camouflaged.

When We Get Back Inside

- Plan to revisit the slash piles to see if animals have been using them. Look for animal tracks leading into and around the piles.
- Research other species in your region that may utilize the shelter to hide from predators.
- New Hampshire Conservation and Natural Resources has created a document on how to create slash piles for wildlife: "NRCS—Creating Brush Piles for Upland Wildlife"
- Vimeo—"Snowshoe Hare Research Video." This is science in action! Students studying wildlife biology and tracking snowshoe hares.

Books about Hares

Summer Coat, Winter Coat: The Story of a Snowshoe Hare—Smithsonian Wild Heritage Collection, by Doe Boyle

Winter's Coming: A Story of Seasonal Change, by Jan Thornhill

Snowshoe Hare's Family—A Smithsonian Northern Wilderness Adventures Early Reader, by Stephanie Smith

Lesson 11. Build a Shelter, Build a Village

Narrative by Pam Dow

2011. Our school year started like none other. On August 28, just two days before our first day of school, parts of our village and town were hit by Tropical Storm Irene. The opening of Moretown Elementary School had to be delayed. Children and families spent the first days after the flood wrapping their heads around the magnitude of the disaster that destroyed their homes and town. Quickly it became neighbors helping neighbors, strangers helping strangers. Although many material possessions were destroyed by the flooding, our sense of community was just beginning to emerge.

Nick Neddo

salvageable that came into contact with the contaminated waters. Although crews worked hard for many days, our school building was still not safe for students. Many of our children were experiencing the most tragic event of their lives. They saw the sides of the main street piled with debris, residents' belongings, and the guts of their homes, not salvageable due to the wrath of the flood.

The children of Moretown needed normalcy. They needed something they knew and loved, which was why the staff knew school must go on. We began school in tents set up on our baseball field.

Many children's homes were flooded, and their possessions were gone or covered in a thick, gooey muck that was left behind after the waters receded. Moretown Elementary School suffered damage as well, and as teachers went about helping families clean out their homes, a professional cleaning crew set to work on the school, ripping up all flooring, scrubbing, and sanitizing anything

With surprising ease, children and adults adapted to our tent school. Teachers taught, children learned. I think we all grasped at this sense of normalcy, this idea of routine, of school. We were together as a class, a school, and a community, and for a few hours we could focus on something other than the flood.

When the time came to return to our classrooms, it was a celebration. The kindergartners were excited to begin a new

Leah Greenberg

quickly took to building with loose parts scavenged in the forest. They built with sticks and stones, leaves, and natural materials. They built and built, from small mouse houses to "sewers" (large holes at the base of trees) that they dug with sticks. The building continued all fall and winter and into the spring. The sticks and stones and their actions represent much more than exploring in nature. These children had been putting actions behind their emotions. They were mimicking the actions of the adults around them, they were rebuilding. As these children watched their town slowly being reconstructed, they too rebuilt their sense of normalcy—rebuilt their sense of safety—while rebuilding their homes in the forest.

2021. For ten years Moretown Elementary School has been giving our students the gift of learning with nature. Our small pre-kindergarten through sixth grade elementary school embarked on the journey of taking children to the forest each week for ECO. Now, a decade later, we once again took to the forest to heal. COVID had arrived on the heels of the cold March winds, like a red-tailed hawk swooping in when least expected. Little did we know that this virus would bring unimaginable changes to our school days and routines. Much like Tropical Storm Irene's devastation, COVID upended life as we knew it. Classrooms stopped being places where children naturally collaborated, as teachers and staff worked to keep children seated at desks spaced at least three feet apart, while frequently reminding children they must keep a safe distance from their classmates. Our forest classroom became our escape, our place

program that connected students with the forest behind our school. It quickly became apparent that this program was vital to the health and well-being of our children. I watched as children eagerly waited for Wednesday mornings to arrive so they could go out into the forest behind our school to explore, build, and connect with the natural world around them. The timing of this nature-immersion experience was perfect; these children were living out the realities of a natural disaster, but were able to escape into the forest for solace each week. This group of five- and six-year-olds we could go to once again to collaborate and move freely. And building with sticks took center stage!

During the month of January our lessons were focused on animal adaptations. The students had remote learning on Wednesdays due to COVID. On a morning meeting via Seesaw, children answered the question, "If you were an animal, what physical adaptation would you want and why?" One student shared, "I would like to blend in because then it would be less likely for a predator to catch me!" The children were tasked with creating an imaginary animal that could survive Vermont's cold winters. They were asked to think about both physical and behavioral adaptations that might help their animal survive. A second-grade student described their imaginary animal, which they called a "White Runner." They shared, "This is a White Runner. Its fur changes from gray in the summer to white in the winter. And it grows thicker fur for the winter."

Back in the classroom on Thursday, our focus question for the day was, "How do adaptations help animals survive cold Vermont winters?" We were in our second week of focusing on animal adaptations, and the children were ready to head to the forest. We prepared to head outside to complete our work for the day. Our outdoor routine typically starts with playing a game tied to our daily focus, and this week the children turned themselves into their imaginary animals they had created the day before. We asked, "How does your animal move? How does it look when it's resting?" Students engaged their imaginations and leapt, ran, and scurried through the

snow embodying their animal. Once we reached our forest classroom, the children were tasked with building their imaginary animal out of sticks and woodland debris. We also suggested creating a shelter for their animal. When all the students were finished, we took a gallery walk through the forest to see and hear about the animals and their adaptations.

Although this was good stuff, as far as teaching and learning goes, it is what happened next in the forest that I find fascinating and worthwhile to share. We have routines that we follow in the forest, and my favorite includes forest choice, after our main activity is completed for the day. On this day, after the gallery walk, children quickly scampered off to engage in their self-directed activities. I moved through the forest, listening and learning from the children, and I noticed a small group of children collaborating and building with sticks. I asked them to tell me about what they are doing. They were building an ambulance and transporting a patient. The patient had COVID, so they were taking precautions to keep the "workers" safe.

They were processing and re-creating their own life, in the woods, with sticks. COVID was real and scary to them, as it is to the adults keeping them safe, but they were choosing to work through their questions, their worries, their hopes by mimicking the adults around them. They were building and processing and healing. Their important ambulance work and building with sticks continued through our entire forest choice time. Their play and their choices were the most valuable lesson that day. They needed each other, the forest, and sticks!

Once again, the forest did not disappoint. The forest opened its loving arms to us, allowing a break from the reality of our new classroom paradigm with a chance to build and to heal. Here, teachers and students had plenty of space to roam freely with no reminders to keep our distance from friends. Instead, we spread out and built, stick by stick, as we made sense of the global pandemic that had consumed us.

Children love to build, and building outdoors took on a whole new meaning during these two life-changing events. These children had time in the forest to heal, to make sense of their lives.

Through their experiences, these children are becoming leaders, eager to assist their friends up a steep hill or share their knowledge of the natural world with others.

Our students have been given the gift of the forest for many years. In 2011 it was the flood and in 2021 the pandemic. Next year, however, it could be a family's divorce, a car accident, or the death of a loved one. I have come to realize that our natural world is a place for healing. Time spent outside is time spent learning about life and about our sense of community. Children need this gift of learning in and with nature, every school year.

Build a Shelter, Build a Village

Grade Level: K–Third

Objectives
- Students will use sticks and natural debris to construct shelters.
- Students will demonstrate creative thinking by considering multiple solutions to a problem and by building prototypes that are tested and redesigned.
- Students will have opportunities to work as a part of a collaborative team.

We Are Exploring These Academic Standards

NGSS.K-2-ETS1-1. Ask questions, make observations, and gather information about a situation people want to change to define a simple problem that can be solved through the development of a new or improved object or tool.

NGSS.K-2-ETS1-2. Develop a simple sketch, drawing, or physical model to illustrate how the shape of an object helps it function as needed to solve a given problem.

What

Whether it is through applying STEM skills, connecting through literacy and math, or supporting artistic expression, we can standardize stick play in many ways. What is not always expected is what happens when students are given an opportunity to immerse in authentic nature play with loose parts.

Sticks are the ultimate open-ended material, and, given space and time, students will construct and deconstruct their world in order to understand it better and their place in it. The social and emotional learning (SEL) that percolates to the surface can be dynamic and meaningful, and we have learned that social skills can be practiced and honed through stick play. The Collaborative for Academic, Social and Emotional Learning defines SEL as simply "a lifelong process of learning how to better understand ourselves, connect with others, and work together to achieve goals and support our com-munities." Students have constant oppor-tunities to practice self-awareness and self-management by building with sticks. Students develop self-efficacy as they follow their interests and feel a sense of purpose when constructing with sticks.

Leah Greenberg

We can also consider how stick play supports the development of transferable skills. Transferable skills are the life skills that can cross over content areas and are demonstrated when students are collaborating, communicating, solving problems, directing their own learning process, and tapping into prior knowl-edge. In chapter 7 of the book *Nature Preschools and Forest Kindergartens*, by David Sobel, called "Risks and Benefits of Nature Play," contributor Ken Finch writes about the benefits of risky play and the potential of learning and develop-ment in five key areas: cognitive, physical, social emotional, creative, and spiritual. The table below shows the benefits of stick play in each of these areas, including language defined by the Vermont Agency of Education on transferable skills, and highlights the domains of SEL created by The Collaborative for Academic, Social and Emotional Learning.

Benefits of Stick Play

Today children have various opportunities to learn, grow, and practice skills in the following areas:

Social/Emotional	Physical/Motor planning
Developing social awareness	Developing spatial awareness
Practicing relationship skills	Engaging large muscle groups
Clear and effective communication	Strengthening proprioception
Perspective-taking	Supports vestibular sense
Supporting collective well-being	Processing small sticks builds fine
Self-management	motor skills
Develop self-control and experience	
self-regulation	
Managing conflict	
Persevering through challenges	
Positive interaction through collaboration	
Making responsible decisions	
Creative expression	**Cognitive/Academic**
Taking responsibility for learning	Critical thinking and reasoning
Developing a sense of agency	Creative and practical problem-solving
Integrating personal interests	Expressing ideas clearly and persuasively
Planning out steps and strategies	Informed and integrative thinking
Positive self-identity through risky play	Apply knowledge and context to real life
Making connections with others to	situations
build skills	Analyze, evaluate, and synthesize information
Feeling and showing empathy for others	from many resources
Celebrating successes and challenges	Develop and use models to explain
Gratitude	phenomena
	Use evidence to justify claims
	Apply mathematical thinking
	Connect to place through natural history
	and the study of trees
	STEAM

How

After proper scaffolding of stick safety, support students to freely build with sticks during a forest choice time. Students may apply what was learned in the focus lesson during their playing. That is always a good sign of a lesson well received! Review your stick safety rules before children begin. Another way to scaffold free choice in an outdoor setting is to have students draw or write about their plan for building with sticks in the classroom. These plans can be brought outdoors and referenced.

Agreements may need to be made on the evolution of shelter building and how shelters are played in and where they are located. To prompt this discussion, ask students what they enjoy about building with sticks and what they find challenging. Be prepared for students to become attached to their structures and protective of who plays with them. With the construction and engineering there are opportunities to learn about territory and valuable resources. Along with the heavy work comes learning how to navigate the social community. How do we include others? Is there room for all? Come back to the three cares and ask, "How we can we care for ourselves and others when we are building with sticks?" You may end up coming up with very specific building or shelter rules. Stick play and shelter building will develop with each season as the materials change with the weather and children progress in their skills. Keep a documentation of students' plans and photos of their creations, and record your observations of how children are learning with sticks. Consider recording your observations using the example above of the benefits of stick play.

We Can Use These Materials Outside

- Bring a camera for documentation of students' work.
- Record children explaining what they are building and what their strategies are.
- Have students bring their stick plans outdoors in a plastic sleeve so they can reference them.
- Laminate pictures of artists who have used sticks and bring them with you for inspiration.
- Keep strong twine, such as jute, in your backpack for supporting structures.

- Have paper and pencil on hand so students can document their work.

When We Get Back Inside

- Discuss successes and challenges of building with sticks that day. Did everyone stay safe?
- Have students share whether they followed their stick plan or not. What worked or didn't work? What would you change about your stick plan? What would you add?
- Write directions to how to build a stick shelter. Include photos, diagrams, or a video.
- Invite another class or learning buddies to come build with you. Students become the teachers!
- Investigate animal-built nests and dens and try to mimic building them to scale. Ask questions, such as, how big is a bald eagle's nest?
- Research how to build primitive survival shelters.
- Learn about traditional Indigenous dwellings of your region.

References

Sobel, David. 2015. *Nature Preschools and Forest Kindergartens: The Handbook for Outdoor Learning*. St. Paul, MN: Redleaf Press.

The Collaborative for Academic, Social and Emotional Learning. https://casel.org/

Teaching with Fire

The Heart of an Outdoor Classroom

Rise Up, O Flame
Rise up, O flame,
By thy lights glowing,
Show to us beauty,
Vision and joy.
—*Christoph Praetorius, c. 1600*

The sounds of crackling and the smell of wood smoke. The dance of a flame coming to life. Fire can light the dark of night, provide warmth, and produce a flame to cook over. Fire is as old as we are, and humans have a biological need for fire in order to survive. In unit 3 we invite you to consider the possibilities for how to bring fire into your outdoor teaching practice. When these lessons are properly scaffolded with safety in the forefront, the routine of building a fire and tending it can create resilience for learning outdoors in the colder months.

Fires allow students to practice caretaking of themselves and others—a type of caretaking that is alert yet tender and infused with awe and respect. When a class constructs a fire together, it builds self-esteem and self-worth. The simple act of sitting around a fire is rooted in security and belonging. You will find safety protocols, science, literacy, math, and social emotional learning in this unit. By inviting fire into the outdoor classroom, we will learn that its magic can safely build culture and unite all people.

Narrative by Emily Carley

My class follows a meandering marked trail along the base of Harrison Preserve's ridgeline and enters Watershed, one of our four gently defined forest camps. We are singing a call-and-response song, called "Welcome In," which has become our routine for gathering in our forested classroom. Children settle their packs on the ground, find seating in the meeting area, and are reminded to get comfortable. I lead the group in three mindful minutes of listening and looking at our surroundings before sharing nature notes around the circle.

For the past few weeks, the purpose of our morning meetings inside the classroom was to introduce fire as a routine component of our lessons. The model of a fire safety circle, a pretend fire constructed out of paper, and a fire tender's supply tote had been in the center of the meeting circles for children to observe and generate questions. We learned how to be safe around the fire by practicing in a familiar and trusted place first: our classroom.

Today, at our Watershed camp, there is great anticipation as students know we will be experiencing our first fire in our outdoor classroom. "Are you feeling ready to practice fire safety outdoors today?" I ask them. With heads nodding and voices declaring yes, I begin removing materials from the fire tender's supply tote and place them in front of me. These include a thick white and blue rope, a foil poultry pan, a box of kitchen matches, a small bag of wood shavings, thinly sliced pieces of kindling, a hot mitt, and a half

gallon of water. I watch and listen to the children observe the items I have taken out of the bag. Children are mentally checking for the visuals I posted during our indoor lesson this morning. I use the rope to make a five-foot-diameter circle and place the foil pan in the center along with the other materials.

In order to continue what we practiced inside the classroom with fire safety, I ask two children at a time to practice walking to and from the meeting space around the fire circle area. We agreed the fire tender may call "emergency" if they notice someone too close to or inside the fire safety circle. This call helps children follow their Stop-Look-Think-Do plan for moving safely in our outdoor classroom with our fire routine included.

I enter the fire safety circle and sit cross-legged. I pick up a handful of wood shavings and place them in the foil poultry pan. I gently lean the thinly sliced kindling (about the size of a pencil) in a teepee-like fashion around the wood shavings. These materials are so dry and small they will light immediately, burning quickly and safely.

Our first fire is small enough that children can see the process and make observations without feeling anxious or overwhelmed. I announce that it is time to light our first fire, and bodies settle

Fire and magic are almost the same. They are both sparkly but only one has real unicorns.

—*Rosa, age seven*

and voices hush. It feels actively quiet, and the attention to the lighting of the match is palpable. I pause and ask the students, "Who gets to use matches?" Their responses are affirming of all we have learned in the classroom before this moment. I hear, "Adults do!" and "Yeah, adults need to be there!"

This brings us back to focus on the agreed responsibility we have in relation to fire. I strike the kitchen match and lean in to light the materials. I place the box of matches back in a plastic bag and then immediately into the fire tender's tote bag. Audible gasps, oohs, and ahhhs fill our circle as the shavings and kindling ignite and grow into a flame. We watch how the materials burn, and the children share observations using all their senses.

"It smells like camping!"

"When does it go out? Why is some of the wood white?"

"I know this, I mean I really know this."

"Ms. Carley, you can light a match really good!"

Children continue to eat their snack and make observations as the fire slowly dwindles. We will not add any more fuel to the fire today. When the flame disappears, I gently pour water from my gallon jug into the foil pan and stir

Amy Butler

the embers with a stick. Today we are practicing being caretakers of ourselves and each other. We are learning with and about fire.

The fire safety circle stays in place as we finish snack and get ready for our workshop time. From this day forward and all through the winter, fire will be a part of our learning routine, it will be the heart of the outdoor classroom.

Welcoming the First Fire

Grade Level: K–Third

Objectives

- Students will share their knowledge and experiences with light, heat, and fire through oral expression.
- Students will understand materials used to build a fire.
- Students will apply knowledge from various disciplines and contexts to real life situations.
- Students will understand safe practices when working with a fire.

We Are Exploring These Academic Standards

NGSS.K-ESS3-1. Use a model to represent the relationship between the needs of different plants or animals (including humans) and the places they live.

NGSS.2-PS1-1. Plan and conduct an investigation to describe and classify different kinds of materials by their observable properties.

CCSS.ELA.Literacy.SL.K.2-3.2. Confirm understanding of a text read aloud or information presented orally or through other media by asking and answering questions about key details and requesting clarification if something is not understood.

What

By teaching fire safety and including fire in the routine of learning outdoors, we are modeling for children that we believe they can be responsible for themselves, each other, and their community. It also supports children to move into taking on a role of being responsible within an activity that is risky and enjoyable. Beginning in the classroom, fire preparation and fire safety can be taught sequentially with developmental appropriateness.

How

The best place to start with introducing fire is to understand the students' previous experiences with fire. Not all students may have an experience with fire, nor will they all have had a positive one. As wildfires increase in numbers and severity, it is crucial we understand our students' stories with the element of fire. Not only can fire be a wonderful teacher of science and facilitator of social-emotional learning, but it can also harm and trigger recent or generational trauma. If your students have a history with fire, it is important you gather this information and their experiences before introducing fire to your outdoor teaching practice.

Begin with a discussion about fire. Ask students about their experience with fire. Be ready to set aside time for stories to be told. Start with sharing about light and heat. Have you ever lit candles for a birthday party or a special occasion? How do our homes stay warm in cold weather? Who has a woodstove or has a fireplace in their home? Has anyone had a campfire before? Have you roasted marshmallows or hot dogs? Ask students what they know about fire and fire safety. Explain that today they are going to create safety rules for building and tending a fire and for enjoying the warmth of a fire.

We Can Use These Materials Outside

- Prepare a fire tender kit with students inside the classroom to bring outdoors. This includes materials to start a fire as well as tools for staying safe.
- Bring a laminated copy of your fire safety checklist (see appendix 9) as a point of review and for guests.
- Have the students build the fire safety circle at your site from the start of your program so that students are aware of it. Students can make the fire safety

area by collecting rocks to make the circle or by using a brightly colored rope and deciding where it can be laid down.

When We Get Back Inside

- Have older students write the fire safety rules and present them as a formal document to be used throughout the school. Teachers may assign students to research how fire was used in the past by other cultures and how we use it now.
- Younger students can draw pictures or create mini models of what safety looks like around a fire.
- Create a "fire unit book" about fire, both from the class's time outdoors and from their families and greater community.
- Invite your local fire department to be a part of your fire safety lessons. How can they support the learning?
- Take a walking field trip to your school's kitchen to see how heat is used to cook breakfast and lunch.
- Think about the cultural connections that can be made to light, heat, and fire in your community.

Susan Koch

- Research the Indigenous knowledge about fire in your region. How does traditional fire knowledge support biodiversity?

Fire Safety Scaffolding Inside the Classroom

- Have students build a pretend fire in the center of your morning meeting area. Include a fire ring to hold the fire.
- Use a thick rope or brightly colored yarn to create a Fire Safety Circle.
- Practice walking around the OUTSIDE of the fire area.
- Generate students' ideas on how to be safe around a fire. Begin with: What do we know about fire? What do we need to be careful of? How will we stay safe?
- Create and agree upon fire safety rules (see appendix 9).
- Create a social story of your fire safety rules. Laminate and bring outdoors.

Ms. Carley's fire tender tote bag

Much of the joy educators experience in the moment with children is a result of time well spent preparing for learning. In the narrative I shared above, I used a fire tender's tote bag, which was full of materials I needed to safely start and extinguish a fire in our outdoor classroom.

I created the fire tender's tote bag to meet my own need to manage fire-making materials and have them all in one place, ready to go. The organizational approach also supports my own hope for immersion in the soulful routine of making and enjoying fire with my students in our outdoor classroom.

All supplies I included in the tote are materials that are readily available at local grocery and hardware stores, so it is easy to both create and replenish. The tote is most useful if it has two sturdy long loop handles. It can be a reusable plastic bag or made of rugged canvas. I do not store fire making supplies on site in our outdoor learning space, so I pack my tote with enough fire-starting materials, including kindling and cordwood, for a single outing; I replenish these weekly. To eliminate the risk of dampness, I keep any paper-based materials and my matches in sealed plastic bags.

Supply list

Two aluminum foil roasting pans (21 × 13 × 3 inches)

One oven mitt

One fire safety circle rope

One half gallon of water

One set of adult work gloves

Hand spade shovel

Box of single-strike matches, flint and steel, and a lighter

Shredded or rolled newspaper

Northern white cedar bark, paper birch bark, or other native flora for tinder

Kindling, thinly split

Fire-extinguishing blanket

First aid for burns

Lesson 13. The Math in Gathering Firewood

Narrative by Emily Carley and Amy Butler

A week after our first fire during ECO, back in the classroom, we reflect as a group about our shared experience. This process of reflecting on what we learned helps the students make more connections and supports them in our next task, which is building a fire as a

"We don't run near the fire. We don't want to fall."

Our getting-ready routine is really dialed in now, and students are taking responsibility for themselves by getting dressed for a morning in the forest on a chilly day. Students look at the clothing

Nick Neddo

class. We discuss what we noticed about our first fire and how we felt about it. By recalling our outing from last week, and by tapping into our emotional experience, we are sure to get everyone remembering the fun we had and also the safety we practiced. Even inside the classroom, the reverence and awe about having a fire in the forest comes rushing back.

"Only the fire tender gets to go inside the fire safety circle."

"Adults light the matches"

"This week we need to sing our fire song!"

checklist for the day and remind each other to pull their snow skirts over their boots and "hoods are not hats," a reminder we created together as a group. There is an impressive level of orderliness in preparing to leave the building, and soon we are on our way!

We again reach Harrison Preserve and sink into the forest in our meeting area. Our game for the morning will be about collecting firewood, so the students are seated in a circle around our new fire area. The children anticipate a fire again as we did last week, and voices are quiet as I start to pull the ingredients for making a fire from the fire tender bag. I ask the students, "Are all the materials here for having a fire and for practicing safety?"

After a pause and a settling of chatter a student asks, "Yes . . . but where's the firewood, Ms. Carley?" In response, I pull from my pack several color copies of three animal photographs and place them on the ground around the circle. I suggest to the class that these animals may help us prepare for fire, and the I ask the students what they know about the animals that are pictured: mice, chickadees, and beavers.

Students begin to share ideas and the naming of animal attributes begin to fly around the circle. I listen intently. Some of the children think beavers are able to carry trees, and this idea is quickly reduced to sticks about the size of a first grader's arm, which we call beaver sticks. Mice have long, thin tails, and it's agreed that sticks this size, which we name mouse tails, are good fire starters when dry. The black-capped chickadee, which is native to Vermont, is a light bird that can land on branches that are between the sizes of mouse tails and beaver sticks. To wrap up the thinking about these animals, I share a story called "Four Animals and the Fantastic Soup," about how four animals help a child start a fire in order to make a pot of soup. I let the children know this wonderful story was written by our friend and the naturalist-educator we call White-tailed Deer, but who is more formally known as Carrie Riker.

Four Animals and the Fantastic Soup

Story by Carrie Riker

There once was a child who, all summer long, had been daydreaming of making soup from their garden. They grew spinach, carrots, potatoes, beets, onions, broccoli, and kale. The perfect fall day for soup had finally arrived, and so the child got a big soup pot, some fresh water, a knife, and a cutting board. They gathered all the harvested vegetables and headed out to the fire pit.

The child chopped everything up and put it in the pot. They were ready to make soup—but first, they needed a fire! The child put the big log they had into the fire safety circle area and struck a match. They held it up to the log and "pfft!" it went right out. They tried again. "Pfft!" The log wouldn't catch on fire. The child was sad—were they going to have to eat cold soup?

The child sat there thinking so hard about this problem that they soon dozed off in the afternoon sun and fell asleep. While they were sleeping, a little creature came up and gently tapped them on the leg. The child woke up to see a woodland jumping mouse sitting there!

"Hello," said the mouse. "I'm here to help you make a fire. I've gone to fire building school and there's a special type of stick you need to use to start your fire. You should look for sticks that are just the size of a mouse's tail, like my tail! Do you see how long and thin it is? You want to find some that make a snapping sound when you break them. You will need many, many, many mouse tails to start a fire. So, look closely on the ground to find the ones that have fallen from the trees."

"Oh," said the child, "I didn't know! Thanks so much! By the way, would you like to stay for some fantastic soup, woodland jumping mouse?"

"Oh no, thank you," said the mouse. "I don't care for the taste of soup! I like to eat insects and seeds. But good luck with your fire!"

The child gathered many mouse-tail-sized sticks and was about ready to try to light the fire again when a small bird landed nearby in a tree. This small bird had a cap of black feathers. It was a black-capped chickadee.

"Hello there! Are you trying to make a fire?" asked the chickadee. "Yes," said the child. "I just got a tip from a mouse about gathering mouse tails and was going to light my match and get it going."

"Well, I have another suggestion for you," said chickadee. "There is another type of stick that you might want to gather, I like to call them chickadee sticks. These sticks are about the size of the stick I'm sitting on right now. Perfect for perching, as I always say!"

"Aha!" said the child. "I will gather those next! By the way, chickadee, would you like to stay and join me for some fantastic tasting soup?"

"Oh no thanks, I don't really like soup, I prefer seeds and berries. Today I am mostly interested in eating invertebrates. Good luck with your fire!"

The child gathered some chickadee sticks, and just as she was putting them in a pile on top of the mouse tails, they heard another noise of an animal moving through the forest. There, waddling through the woods, was a beaver.

"Wait, wait!" said the beaver. "You need something else for that fire! Don't forget the beaver sticks!"

"Beaver sticks?" asked the child.

"Yep!" said the beaver. "Sticks like I would use to make my dam or lodge. Definitely larger than those other sticks. The roof of my lodge would cave in if it were built out of those mouse tails! Make a nice pile of them so you are ready when your fire gets going!"

"Oh, thank you so much!" exclaimed the child. "I am learning many things today. And would you like to join me for my fantastic soup, beaver?"

"Nope, not a soup fan!" answered the beaver. "I am strictly an herbivore, and today I am eating willow saplings. But thanks! Have a good day!" With that, the beaver waddled away.

The child gathered a bunch of beaver sticks and was now ready. Just as they were about to light the match, the child stopped for a moment to gather their thoughts and reflect on all the hard work they had done . . . when just then, a super-soft thumping sound came through the woods. It sounded like soft furry feet hopping about. The child looked around, and there was an eastern cottontail rabbit! The rabbit came very close and didn't say a word. Its nose twitched and its eyes blinked just once.

"Hello Cottontail Rabbit," the child said quietly and cautiously, so as not to scare the rabbit away. "I am thinking that you, perhaps, have something to tell me about making a fire?"

"Well, as a matter of fact, yes," said the rabbit. "I see you could use one more special thing, something soft to make a nest on the inside of your fire structure. Either paper birch bark you found on the ground, or perhaps some cedar bark, all fluffy and soft. Sort of like my tail!"

"Wow," reflected the child aloud. "I have learned so many things today about making a fire. Would you like to stay and join me for some fantastic soup, by the way?"

The rabbit said, "Yes, I love fantastic soup! It is my favorite kind of soup!"

And so, the child gathered some birch bark for tinder and began to build a fire structure that looked like the shelters they had built with friends. First, they started with the mouse tails, then they added chickadee sticks, and then the beaver sticks on the outside. Cottontail rabbit watched closely as the child gingerly placed the bundle of birch bark inside the fire structure and lit a match. The flame touched the tinder, and the fire came to life, the flames lighting the three different sizes of firewood.

The child cooked the soup, poured it into two bowls, and the rabbit and the child enjoyed it together. And guess what? The soup tasted fantastic!

After the telling of "Four Animals and the Fantastic Soup," the children are eager to begin collecting materials for our first fire. In addition to learning about the sizes of the firewood from our animal friends, we also learned that our materials need to be dead, down, and dry. These three Ds supported our third care of taking care of the environment and guided the children to be aware of not taking branches from living trees. Only materials we can find on the forest floor that are not attached to a tree are allowed to be gathered. Finding dry wood is an extra bonus on this damp winter day.

At our central meeting space, we lay out the photographs of the woodland jumping mouse, the black-capped chickadee, and the beaver. I request they match the size of the wood they find with the picture of the animal. This approach engages students in sorting and classifying. Some children head toward parts of the forest that have larger sticks resting on logs. A few students find a stand of pines and source twigs that are mouse-tail sized. Three armloads of chickadee sticks are returned by a small group of quick-footed children.

Twenty minutes pass to the sounds of bodies in motion attuned with intentional forest work. I call the group back together from the reaches of our boundaries with the call of a crow, "Caw, Caw!" We gather in our meeting space to assess the firewood collection; students reflect and compare the piles to the ingredients of our indoor model fire from the morning meeting. We agree we have what we need to make a fire.

For this fire I wait for the group to shift into stillness and ask for silence before beginning to build. "This will be a time for you to listen to the fire," I explain. "Today's fire will be bigger and burn longer than last week's fire. We will talk all about what you hear and see after. I notice your whole body listening. This feels safe because everyone is still, thank you for caring for yourself. Let's get started!" With that, I enter the fire safety circle with all three types of firewood sizes at the ready.

Once the fire is lit, I pause before beginning to sing our fire song. The children join me in singing, and the forest decompresses around us. We process the sounds and sights of fire-making together. On cue, students pull a snack out from their packs and settle in to watching the fire, eating, and visiting with the students close to them in our circle. Again, the same as last week, I ask two children at a time to practice walking to and from the meeting space around the fire circle. We will continue practicing fire safety as a routine to build on our awareness and muscle memory of safety.

At the close of the lesson, I find myself reflecting and feeling deep appreciation. The group disperses to their workshop learning spaces, but one first grader quietly joins me in the fire safety circle. In a near whisper, he cautiously asks if one day he might be the fire tender's helper.

"I already know how to chop firewood into little pieces," he explains. "I've done it with my dad at his tent house when we hunted squirrels. I want to be a firefighter one day." Holding this child's earnest plea in my heart, I realize he is requesting that I come to know and understand him. This new trust and human connection are also the magic of fire.

Karen Cingiser

The Math in Gathering Firewood

Grade Level: K–Third

Objectives

- Students will explore materials used to build a fire.
- Students will practice low-impact harvesting methods for gathering firewood.
- Students will sort and categorize natural materials for fire building.
- Students will identify ways in which they take care of or hurt the environment.

We Are Exploring These Academic Standards

NGSS.K-2-ETS1-2. Develop a simple sketch, drawing, or physical model to illustrate how the shape of an object helps it function as needed to solve a given problem.

NGSS.2-PS1-1. Plan and conduct an investigation to describe and classify different kinds of materials by their observable properties.

NGSS.3-5-ETS1-2. Generate and compare multiple possible solutions to a problem based on how well each is likely to meet the criteria and constraints of the problem.

CCSS.MATH.CONTENT.K.MD.B.3. Classify objects into given categories; count the numbers of objects in each category and sort the categories by count.

What

After practicing and establishing safety routines within the context of having a first fire, we can now include children in the process of gathering materials for a fire. This is an active way for students to move throughout the outdoor classroom space and learn about natural resources that can be used for fire building. The story "The Four Animals and the Fantastic Soup" brings together fantasy, imagination, and the sequencing of a successful fire structure. As children look for mouse tails, chickadee sticks, and beaver sticks, they are learning about the landscape in a new way by mapping resources in the outdoor classroom. The gathering of fire-making materials also instructs children about ethical and low-impact harvesting techniques.

How

There are a few ways to approach this lesson, and the story of "The Four Animals and the Fantastic Soup" is a great springboard. The lesson can also be introduced by having the materials for fire building (mouse tails, chickadee sticks, beaver sticks, matches, flint and steel, and tinder bundle) ready for students to explore.

These materials can be introduced before the story or after. Set the materials up on tables to be explored in small groups, or do it as a class in a morning circle in your outdoor classroom. Have students use their sense of touch and smell to explore these materials. Students can write down descriptive words for all three types of firewood.

Next, discuss the different things needed to start a fire (fuel, oxygen, and ignition). Invite students to explain how they would use these materials to start a fire. Is there a specific order that would work in making a fire structure? What materials would bring the quickest flame? The slowest? Are the materials dry, damp, or wet? What is the purpose of a tinder bundle, and what materials are well suited for making one? An extension for older students would be to learn about the differences between deciduous and coniferous materials and how they contribute to fuel for a fire.

After this hands-on exploration and discussion, students can then go out and collect all three types of sticks for the fire and return to the circle. If you do not have access to a forest with abundant materials, this lesson can be staged with materials by splitting dry kindling to the three sizes. These prepared materials can be spread throughout the outdoor learning area and students can look for them much like a scavenger hunt!

Remind students about the three Ds: dead, down, and dry. How long do the materials need to be? Remember the Golden Stick lesson in unit 2? Use this prior knowledge to support this lesson! How much should students gather? This depends on the purpose of the fire and how long you plan to be outdoors. And always remember, gathered firewood is a wonderful resource to have. Cover it up and keep it dry with a small tarp and tuck it away for next time!

We Can Use These Materials Outside

- Bring laminated pictures of the three animals from the story. Lay these pictures out for helping to sort the stick sizes.
- Make popsicle sticks labeled with the words mouse tails, chickadee sticks, and beaver sticks. Put all the popsicle sticks together in a bag and have students each draw one from the bag. Their selection is the type of firewood they will look for. You can ask the students, "Do we need a lot of mouse tails today because it has been rainy? How many beaver sticks do we need?"
- Add numbers into the fire lesson by using dice, playing cards, or stones with numbers written on them. These can used to dictate how many sticks students need to gather.
- Bring rubber bands for making chickadee stick bundles. Have students make bundles of five or ten sticks.

Amy Butler

When We Get Back Inside

- Think about what other types of animals could be used in the story "The Four Animals and the Fantastic Soup." What animals live in your region that could help tell the story?
- Learn about native tree species in your region and the difference between deciduous and coniferous trees. Which species make the best fuel for fires?
- Learn about how humans started fires before the invention of matches. Connect with people in your community who practice primitive fire-making skills.

Lesson 14. Bringing It to a Boil: Introducing the Kelly Kettle

Narrative by Amy Butler

There are classes of children we never forget. The ones who challenge us and make us question our ability to teach. They show us all the cracks in our routines and have us shaking our heads in humor or frustration. Years later, I still remember these ones. In fact, one of them taught me the power of teaching with fire.

The kindergarteners in this particular class were inquisitive, active, and rambunctious. They loved each other hard and argued and fought just the same. There were so many strong and vibrant personalities! Many of the children were also coming to school daily with the trauma of life at home in their bodies and spirits. As any educator can imagine, bringing your challenging school year into the wilds of the outdoors is a whole other level of teaching.

My co-teacher and I quickly found that settling into our learning routines in the outdoor classroom was taking a bit more time than expected. Each week we had to do some extra planning and preparedness in order to keep one step ahead of them when it came to safety.

Our class would review the Three Cares each week before heading out to the forest, but these agreements seemed to wear off by the time we were under the branches of the hemlock trees. Sitting in circle for snack and story was especially hard as they were not only stimulated by the surrounding environment but also distracted by each other socially. There was arguing over who sat where and constant interruptions during story. Children sprung up to leave the circle, leaving their snack refuse on the ground.

Nick Neddo

This new weekly experience highlighted their abilities as five-year-olds and the places we wanted to support their growth. Inside the classroom we were working on being kind communicators, asking for help to solve problems, and

Amy Butler

taking responsibility for our personal decisions and actions. In the forest we were expecting children to do the same. The forest, I believe, challenged them in a way they had never known, and they were trying to understand how they felt in an unbound space.

My co-teacher and I took careful steps to support the children. We made small incremental steps toward spending more time outdoors. We held the routines tightly and decided to end our time before we spiraled out. As teachers, we were exhausted at the end of it. This, we knew, was not the baseline we had hoped for.

Then, the end of October drew near, and the weather got colder. I made a radical suggestion to my teaching partner: "I think we should introduce fire via the Kelly Kettle." The Kelly Kettle is a portable stove that would allow us to safely have a fire and boil water for tea in our outdoor classroom. Still, my partner looked at me shocked and responded, "You're kidding right? We can't even get through a story without some major redirecting and reminders."

But I persisted: "I'm serious," I said. "I think it could ground them not only in their fascination, but also by including them in some real responsibility that involves a level of risk. We know they are big-time risk takers by how they play. We know they are capable of holding one another accountable for behavior, even if they provoked it in each other. I want to try it."

My teaching partner agreed, and I thanked her for being ready, willing, and able—and brave. And for believing in the capability of these children. In the end, our idea paid off and we were met with great success.

We took the next three weeks to scaffold the idea of fire safety through sharing our experiences with fire and practicing the fire safety circle inside the classroom and in the forest. And, because my co-teacher was also an art teacher, the children made their very own teacups. This was incredibly special, and they took great pride in their cups and could not wait to use them!

On our Kelly Kettle Day, we introduced the kettle inside the classroom first. We looked closely at the parts and noticed how it smelled like a fire. All the children were stunned to see how it worked. They asked a lot of questions.

"Is it magic?!"

"It's a magic kettle! The water will be hot even with the hole!"

"Let's use it! We want to collect the mouse tails and chickadee sticks!"

We geared up and headed to the forest with our teacups and the Kelly Kettle filled with water. Two students helped to carry the fire tender bag. There was a sense of purpose on this day and whole class participation was high. Students placed their backpacks and cups at their spots in the circle and dashed off to gather their mouse tails and chickadee sticks. After ten minutes we called them in. Fire materials were sorted and piled near the fire tender's supply bag, and children settled into circle and got out snacks. Today the story for snack was being told by the kettle and the fire—no words from the adults were needed. We explained to the children that in order to really hear the Kelly Kettle's story, it would be important to practice our fire manners: quiet bodies, quiet voices. We told them, "Today during story, you can use your deer ears, owl eyes, and bear nose. Listen, watch, and smell."

We moved through the steps of securing the kettle in the fire safety circle and began to add the materials collected to the fire bowl that acts as the kettle's base. As I held up the matches to signal that the kettle was about to be lit, children hushed each other and brought attention to the center of our circle.

"She's about to light it. We need to watch!"

"Hey! Everybody, listen! It's time!"

With one match strike, the kettle was lit, and the crackling of the paper birch and mouse tails could be heard. Smoke began to rise up through the chimney, and soon a flame danced out of the top. Children began to speak in strong whispers, and bodies vibrated with excitement but stayed safely in place. Doing their best to keep their voices and bodies quiet, they shared their reactions in loud whispers.

"I see the smoke!"

"Look it's fire coming out of the top!"

"Is the water hot? How does it boil?"

As the water heated up, we drew attention to the water chamber on the kettle and directed children to use their "owl" eyes to look for a sign that the water was boiling. We talked about looking for steam, which would be different from the smoke from the fire. With a lively fire, we had hot water for tea in about four minutes. Children saw the steam and we knew the water was ready.

Slipping on a hot mitt we lifted the kettle off the base of the stove and set it on the ground in the fire pit. Teacups were collected and lined up at the edge of the fire safety circle. The class had decided on peppermint as their first tea flavor. Tea bags were dropped into cups, and the hot water was poured. The adults passed out the tea and children chattered and held their warm cups with the utmost care.

The memory of this day sticks with me because these kindergarteners demonstrated that fire is a teacher. A teacher of stillness and connection. Of care and respect. The Kelly Kettle became a weekly routine for this class. It grounded them to place and to each other. I believe it gave them a sense of how to take on an otherwise adult role and how to practice awareness of self in an unbound space.

Bringing It to a Boil—Introducing the Kelly Kettle

Grade Level: K–Third

Objectives

- Students make informed, healthy choices that positively affect the health, safety, and well-being of themselves and others.
- Students will understand materials used to build a fire.
- Students will practice the fire safety protocol.

We Are Exploring These Academic Standards

NGSS.K-2-ETS1-1. Ask questions, make observations, and gather information about a situation people want to change to define a simple problem that can be solved through the development of a new or improved object or tool

NGSS.2-PS1-4. Construct an argument with evidence that some changes caused by heating or cooling can be reversed and some cannot.

NGSS.4-PS3-2. Make observations to provide evidence that energy can be transferred from place to place by sound, light, heat, and electric currents.

What

The Kelly Kettle is a wonderful tool that allows you to learn with fire in a safe and contained way that also leaves no trace. A Kelly Kettle can be used on a playground, in a school parking lot, or in a city park (all these places with permission of course!), so they make a lot of sense for programs that have limited outdoor space for learning with fire. Kelly Kettles are easily portable and don't take up space to store in the classroom, either. There are immediate gratifying results of using a Kelly Kettle such as hot water to make tea, a bowl of hot embers to roast an apple over, or an opportunity to learn how to safely tend a small, contained fire. Whether you have access to a forest or a park or your school yard, the Kelly Kettle is accessible to all students of outdoor learning.

Susan Koch

How

Before introducing the Kelly Kettle to your students, be sure to take time to learn how to use it! Practice at home or with other teachers, and get efficient with the kettle. Learn about the types of fuel you would find in your area. If you don't have access to natural materials, you can use wood shavings and kindling that is thinly split to size.

Introducing the kettle for the first time can be turned into a mystery to solve for your students. Allow students to explore the pieces of the kettle first and make inferences about it. What do they notice and how do they think it might work? Does it remind them of anything?

Prior to using the Kelly Kettle, students should have already practiced fire safety in the classroom and discussed how they can be safe when near a fire. The Kelly Kettle may be your first introduction to fire. If using the kettle is the entry point for fire safety, follow the scaffolding for safety from lesson 12. Review the fire safety protocol in appendix 9.

Setting up a safe area for the Kelly Kettle is the same as in lesson 12. Remove debris from the forest floor in the area in which the Kelly Kettle will sit, and set up a large safety circle with sticks, rocks, or leaves that only the fire tender (teacher) may enter. If the kettle is being used in the schoolyard or in a parking lot, the safety circle can be made with a brightly colored rope or with chalk

> I came to appreciate that task of tending fire really requires one to cultivate tenderness—attentive, patient, alert and observant. In essence, being with fire was quite similar to how I strive to be with my students.
>
> —*Lindsey Vandal*

on the ground. Igniting the kettle can be done with prepared materials, or you can follow lesson 13 on gathering firewood for the kettle. The Kelly Kettle only needs mouse tails and chickadee sticks for fuel. Dry pinecones and grasses can work too.

Before lighting the kettle, confirm that everyone is sitting with quiet bodies outside the safety circle, and that all other fire safety protocols are being followed. If all safety procedures are being followed, light the fire. If this is not the first fire, you can designate a fire helper to hand you sticks to feed into the kettle.

Have students watch smoke rise out of the top, and also have students watch for steam starting to rise from the mouth of the kettle, as that is a sign that the water is hot enough to make tea. Ask students to think about questions such as, What is the steam? Where did it come from? Why does water do that? Pour water into cups or mugs at a designated tea-making station within the boundary of the safety circle, and hand out tea to students once it has steeped. Use extra water to extinguish the fire by pouring it over the hot coals in the Kelly Kettle's fire bowl, then stir the water and coals to put the fire out. Allow time for all pieces of the kettle to cool before packing it up.

We Can Use These Materials Outside

- Teachers can purchase a Kelly Kettle online at www.KellyKettle.com
- Bring your fire tender tote bag from lesson 12.
- Bring your laminated pictures of mouse and chickadee for helping to sort the wood sizes.
- Again, you can use dice, playing cards, or stones with numbers written on them to dictate how many sticks students need to gather. See lesson 13 for more on this!
- Bring yummy tea and reusable teacups for making and drinking tea.

When We Get Inside

- Students can learn about the history of the Kelly Kettle online at www. KellyKettle.com.
- See the upcoming lesson 15 for books about tea.
- Learn a native legend about fire, such as *Fire Race: A Karuk Coyote Tale of How Fire Came to the People*, by Jonathan London. Support students in discussing cooperation versus stealing in relation to this legend.

Lesson 15. Teatime with Trees

Narrative by Susan Koch

"It's teatime!"

"Teatime everyone!"

The twitter of the first graders resonates through the forest on this mid-February morning. The class gathers in their base-camp, which we have named Fern Hill. Some children remove their snowshoes before entering the circle. Others have laid down their snow babies, which are little make-believe babies the students made out of snow and sticks, in order to be fully present for teatime. The snow is so deep that it is difficult to find the stumps placed in a circle, but they manage. Children pull mugs from their backpacks, and some even put a bit of snow into their mugs, as they know the tea may be hot at the first pour.

As they await their tea, the chatter resumes.

"My hands are so cold; I can't wait for tea!"

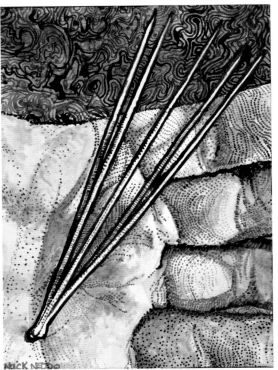

Nick Neddo

"I love teatime, teatime, teatime, teatime!"

"It smells like my grandma's house."

"I hope I get bonus in my tea." (This child is talking about pine needles or tea leaves in their cup.)

"Does everyone have some? I can't wait!"

This routine of preparing and tasting tea together has become a class favorite. This week the class is tasting their first homemade tea out of the needles from eastern white pine trees. Last week the class harvested white pine boughs in our outdoor classroom. Then, we brought them back to school and explored them during the week in our nature museum, which is a part of our classroom. Students easily recalled the trick for identifying eastern white pine trees: remembering there are five letters in white and also five bundled needles on a white pine.

The white pine boughs were added to

Harvesting ethics

Please note the processes outlined in appendix 11, ECO Edible Plant Protocol for Making Tea. It is our responsibility to avoid harvesting endangered or toxic plants. By practicing the Honorable Harvest, as outlined by author and Potawatomi citizen Robin Wall Kimmerer, we can learn how to be in reciprocity with nature. Teaching students about plants demands that we, the adults, understand beyond a shadow of a doubt how, when, where, and what to harvest. Invite a local plant expert to teach you about three local, common, and safe plant species you can use with your students for making tea.

plants that students have harvested and processed is a special activity. We have noticed some children who may be considered picky eaters are apt to take a risk in trying something new because they were a part of the preparation. The feeling of pride and purpose can be seen when children are serving tea to one another. In the act of collectively sipping tea, students can express their sense of taste with descriptive vocabulary terms. We can encourage students to expand their experience beyond "I like it" or "I don't like it." When tea becomes a routine of the outdoor classroom, students have time to get used to it and to try the tea over the course of the school year. It can be an invitation to step out of their comfort zone in a seasonal time frame that is suited to the individual student.

We Can Use These Materials Outside

- Sometimes we do not have time for a fire or are not able to have one. That's okay! A stainless-steel vacuum insulated thermos will keep tea warm and is easy to bring along on an outing. Coleman and Stanley both make great 40- to 68-ounce thermoses.
- Purchase teacups for your classroom to use outdoors. These will get a lot of wear and tear, so consider stainless steel. Some enamel type mugs can chip paint. High quality mugs will last you a long time!
- Use a Kelly Kettle if you are able, and don't forget your fire tender tote bag.
- A mortar and pestle is a great tool to use outdoors or in the classroom for processing plants for tea such as pine needles and woody herbs.

I love the combination of the warm tea in a little cup and sharing its energy with your hands . . . all the happiness it brings to the tongue and belly! Smiles! Perhaps there is a scrunched-up face or two from those who decide it is not quite their thing and they just carefully pour it out, returning it to the earth. I have seen students choose a plant for tea that they have previously adopted, promising to regularly check on it throughout the coming year. There is a powerful sense of self-efficacy to have children physically engaged in taking care of themselves by making a warm drink.

—*Carrie Riker, teacher-naturalist,*
North Branch Nature Center

Hygge (noun): a quality of coziness and comfortable conviviality that engenders a feeling of contentment or well-being (regarded as a defining characteristic of Danish culture).

Back Inside the Classroom

- Research native plants in your region that you can safely use as tea.
- Work with local garden enthusiasts to establish a tea garden at your school.
- Create a space in the classroom where students can make their own tea. This can include an electric teapot, teacups and saucers, and boxes of herbal tea such as chamomile and peppermint. What would it be like to have a daily teatime in the afternoon during read aloud or a silent and solo time?
- Create a sensory word bank at tea tasting times. Do the words we use to describe peppermint tea sound the same as the words we use to describe white pine needle tea?

Books about Tea Drinking

Tea Party in the Woods, by Akiko Miyakoshi

Teatime Around the World, by Denysse Waissbluth and Chelsea O'Byrne

African Tea, by Carter Smith, illustrated by Lesley Tomsett

Cloud Tea Monkeys, by Mal Peet and Elspeth Graham, illustrated by Juan Wijngaard

A Kid's Herb Book: For Children of All Ages, by Lesley Tierra

Lesson 16. Cooking on a Stick

Narrative by Amy Butler

Students carry a bundle of three-foot-long greenwood sticks up into their basecamp behind the school. This established outdoor learning space is in the town forest that abuts the school property. As convenient as this is for a wonderful learning space full of affordances, we still need to climb one ridiculously steep hill a community of hunters, farmers, and multi-generational Vermonters who highly value the outdoors. Needless to say, going to the forest every week, building shelters, identifying trees for tea, and learning fire safety is embraced by parents and caregivers in this community.

On this particular morning, the two

Nick Neddo

and then another hill that winds into the forest. At this point in the school year the children have built stamina for climbing and a love for their basecamp, and all that happens there. The teachers have adapted by knowing who needs some extra help up the hills and where they themselves need to stop for a breath!

At this school we are lucky to have a grandparent and a parent volunteer with us every week. This is not typical, as the demographics of the surrounding community are mostly full-time working families. They are also largely volunteers were already in the forest and had started a fire. This was especially helpful because, today, we would be cooking over the fire. This was a culmination of our fire safety lessons and a wonderful way to celebrate the upcoming school break and to share food together as a class. The fire got started early because we needed a good bed of coals to successfully cook a favorite recipe, bread on a stick. As we continued the climb, questions were coming fast, and bodies were moving just as quickly.

"What are those sticks for?"

Amy Butler

"Are we having a fire today?"

"Those are marshmallows sticks! Right?! Right?!!"

"Who is the fire tender?"

"What are we doing today?"

"Hey! I smell the fire! Who's in our base camp?"

Children tumbled into the basecamp, and they placed their backpacks near the backpack tree and found a seat around the fire. Chatter continued and the children settled in, as they chatted with our beloved adult volunteers. Of course, they both got hugs from their own children, but many of the other children hugged them and thanked them for the cozy fire. This was a sign of a connected community of learners. The goal is that we work toward the classroom teacher not being "in charge" all the time. The environment, the adult-to-child ratio, and the routines of learning with nature hold everyone.

The first activity of the day was to whittle our roasting sticks. In the classroom we had reviewed safe use of these high-grade potato peelers by teaching the students the Whittling Peeler Safety Protocol (see appendix 12). Practicing indoors looked like modeling the protocol with a pencil and a marker, the pencil representing the whittling tool and the marker representing the stick. Students tried sitting safely in an upright position, setting up their safety circle, and repeating the movement of placing the peeler on a stick and whittling away from themselves. The amount of focus in the classroom was so high! The children knew that the next day they would use these real tools to whittle their very own roasting stick for cooking over the fire with.

In the forest we set up an area where eight children at a time could whittle their roasting sticks. One adult was present in the tool area. One adult was present at the fire (always!) and two adults were available to support free choice time.

There was a noticeable silence of focus in the tool area. Children felt purpose and importance in the task of whittling these roasting sticks. If we had not been using them that day to cook food, it was apparent that students would have whittled their sticks down to a toothpick!

As students finished whittling their roasting sticks, they were invited over to the fire to cook either an apple slice or some bread dough. Children sat closer to the fire than they had before, and this created a noticeable sense of responsibility. The ratio was five children to one adult at the fire. Children roasting bread sat patiently and watched the dough expand and brown.

"I can smell it!"

"Ohhh! It smells soooo good!"

The children noticed that the roasting apples spit and oozed as the sugars cooked. The finishing touch was a shaker container of brown sugar and cinnamon presented by the classroom teacher. "When your treat is cooked how you want it, come see me and we will add something extra," the teacher explained.

In the rest of the outdoor classroom, children whittled in the tool area, reconstructed mouse houses, and gathered beaver sticks for the fire. There is this quiet buzz we refer to in teaching outdoors. It happens when the routines have been embodied and each student feels it, head to toe. This buzz is the sound of aliveness and autonomy, of self-confidence and generosity to others. This is what we have been working toward. Within this quiet buzz, I see the classroom teacher lean back against a tree and sprinkle cinnamon sugar on bread and apples for their students. A relaxed smile spreads across their face, and the buzz continues to reverberate through the forest.

Cooking on a Stick

Grade Level: K–Third

Objectives

- Students will display safe practices around a fire.
- Students make informed, healthy choices that positively affect the health, safety, and well-being of themselves and others.
- Students use scientific methods to describe, investigate, explain phenomena, and raise questions

We Are Exploring These Academic Standards

NGSS.K-2-ETS1-1. Ask questions, make observations, and gather information about a situation people want to change to define a simple problem that can be solved through the development of a new or improved object or tool.

NGSS.2-PS1-4. Construct an argument with evidence that some changes caused by heating or cooling can be reversed and some cannot.

What

Cooking over a fire is antithetical to how we spend most of our modern life cooking. In our busy lives, it is a rare thing if a family sits around the table to share dinner more than a couple times a week (if at all!). We are a culture of convenience and "get it done and on to the next thing." Instead of gazing into the flames of a fire when preparing food or sharing a meal, many people stare into a lit screen in the palm of their hand. This lesson is an opportunity to show another way, one that is ancient and part of our genetic makeup.

There are many benefits to cooking over a fire, and there seems to be a universal feeling that food cooked over a fire is especially tasty and special. Why? Let's start at the beginning of this unit with lesson 12, Welcoming the First Fire. Now we know how to build a fire and how to be around it safely as a class! Building on that, we know that, in order to cook bread or apples, the fire needs to be built, lit, and tended so there is a good bed of coals for cooking.

Preparing a fire for cooking takes time, participation from everyone, and attentive engagement. Students can be a part of this process. Inviting students to cook their own snack requires that they be focused not only mentally but also physically. The core strength it takes to hold, balance, and rotate a cooking stick involves the whole student. This focus results in a level of being emotionally and spiritually satisfied when the snack is done cooking and enjoyed with peers. Is this why food cooked over a fire tastes so good? We think so!

Amy Butler

How

We encourage that fire safety and fire manners be reviewed each and every time fire is planned as a part of your outing. We all need reminders as we get more and more comfortable and adept at being around the fire. The fire protocol is to be upheld at all times, and children can help us do this. We also know that some school days bring different challenges and needs. By reviewing our safety protocols, it allows us to check in with where our students (and adults) are at, to see whether anyone needs some extra support. Or possibly, that today is not a day for a fire. Sometimes the adults are not prepared, and that is okay. We can all try again next week!

For this lesson, the Whittling Peeler Safety Protocol will also need to be reviewed and practiced before the lesson (see appendix 12). If you do not have access to greenwood for roasting sticks, you can purchase sticks specific for roasting over a campfire. Follow the tips for making your own roasting stick included in appendix 12.

Begin a discussion about cooking over a fire. Has anyone cooked over a campfire before? What did you make? How do we cook food in our homes? How did people cook before electricity? What do we know? What do we wonder? What do we think might surprise us about cooking food over a fire?

In ECO we steer away from marshmallow roasting and offer other fun alternatives that have sustenance and demand focused engagement. Marshmallows cook fast and catch on fire. It's a one and done sugary treat that many children have experienced already, and it can be tricky to safely manage. Even in rural Vermont, not many children have cooked apples or bread over a fire. This is a first-time experience shared by the whole class. Everyone is a beginner together, even the adults!

Recipes

Roasting Apples

Quarter or halve apples, depending on how much time you have to prepare and cook them. You can leave the skin on, as the fire cooks the flesh of the apple underneath, and the sugars will start to caramelize. The skin can be carefully peeled off once it has cooked, but know that the flesh underneath the skin will be extremely hot! Teachers can sprinkle brown sugar and cinnamon on the apple and have students put it back over the fire. Then, get ready for extreme yumminess!

Roasting Bread

Making bread dough is always an affordable and learning-rich option for this lesson. The dough should be made ahead. But also note, premade pizza dough from the grocery store or bread dough from a local bakery works just as well! To make the dough, combine the following ingredients:

- 4 cups flour
- 1 teaspoon sugar
- 1 envelope instant dry yeast
- 2 teaspoons salt
- 1½ cups water, warm

Mix ingredients together and knead firmly for ten minutes. Allow two hours at room temperature for the dough to rise. Transport the dough in a resealable bag. A teacher can gently pull the dough into a long rope-like section and help students to wrap it around their sticks. The "dough snake" should wrap around uniformly in thickness because big dough balls will create uneven cooking.

Modeling how to do this inside the classroom sets children up for more success outdoors. Cooking bread over the fire can take up to fifteen minutes. Demonstrate how to hold the stick over the fire so that it is close to the flame but will not burn. After the dough is cooked, it will feel firm on the outside but will still be springy, and it will look golden to dark brown. Remove the dough from the stick and enjoy plain or with honey, jam, or another favorite condiment.

We Can Use These Materials Outside

- Bring your fire tender tote bag and some kindling and firewood for maintaining a fire long enough to produce coals to cook over. If you can have someone start the fire ahead of time, such as a parent or other volunteer, that would be even better!
- Prepare pre-sliced apples and bring them to the fire in a resealable bag. But, if time is short, apples can be cut up in the forest as needed.
- Make bread dough ahead of time or bring store-bought dough. The bread dough can be already prepared for each child and put into a resealable bag with their name on it.

- Have ready extra resealable or other bags for waste, hand sanitizer, wipes, and disposable gloves.
- To keep sticky hands clean, a thermos of warm soapy water for handwashing is a great thing to have, but don't forget the paper towels!

Back Inside the Classroom

- Bake bread inside. Collect bread recipes from families to try in school or in your outdoor learning space.
- Research the origins of bread and study bread recipes and practices in different cultures. Create a map of where each ingredient came from.
- Draw or write a step-by-step lesson for how to bake bread and roast apples over a fire. Have your students create an instructional video or PowerPoint about how to cook apples and bread over a fire.
- Designate a student-photographer to document the process of cooking. With younger children, take pictures of each of the steps. Print the pictures and have the students put them in the right order for making these treats.
- There are so many other recipes for great snacks to cook over a fire. A simple online search will bring you a plethora of ideas!

Books about Bread

Sun Bread, by Elisa Kelven

Fry Bread: A Native American Family Story, by Kevin Noble Maillard, illustrated by Juana Martinez-Neal

Bread, Bread, Bread: Foods of the World, by Ann Morris, illustrated by Ken Heyman

NICHOLAS NEDDO 2002

Unit 4
Winter Weather, Animals, and Us
Learning Outdoors with Resilience and Wonder

"Can you hold my treasures while I go on a journey?"
A child hands me sumac berries, a goldenrod gall, and a single white-
tailed deer hair. They scale the snowy, slushy hillside, shouting and
declaring new discoveries. Even under gray skies and faced with the
shortest days of the year, children always find joy outdoors. As adults
we might experience this landscape as looming and impenetrable.
Winter can be hard. Although, for a child at nine in the morning
on a school day, it's open, endless, and full of possibilities. It's the
possibility held in a single sumac berry, a goldenrod gall, and a white-
tailed deer hair. Let wonder penetrate the cracks and expand!
—*Amy Butler*

Winter is on its way, with all its wonder, beauty, and challenges of getting out the door with your students. A well-known quote from a famous British author and fell walker, Alfread Wainwright, suggests a simplistic perspective on being outdoors in foul weather: "There is no such thing as bad weather, only unsuitable clothing."

We can allow ourselves to acknowledge the weather is not favorable today and that unsuitable clothing would in fact make it much more miserable. While wild animals are cloaked in thick fur and sleeping in a den, we are struggling with damp winter boots and finding our favorite socks. Yes, teaching outdoors in winter is another level of hard. And the hope of unit 4 is to ease your doubts by learning about more than human animals in winter.

In many climates this seasonal change requires us to develop resilience, find a sense of humor, and unearth adventure. By nurturing awe and appreciation for the living winter landscape, we have a chance to bond with our students that cannot be replicated indoors. Let's seek the benefits gained by creating a learning community outdoors during the coldest months of the year!

Lesson 17. Deep Winter: Caring for Self and Others

Narrative by Jenna Plouffe

It is hard for me to pick a favorite season; I find something to love about and look forward to in each one. But the transition from autumn to winter demands our attention unlike any other time of year. We feel the change in our body. The cold greets our fingers first, closely followed by our ears and toes; finally we feel it in our core. We see the change on the land. There are clues everywhere: the leaves have fallen from the trees, chipmunks are busy gathering food to store for the winter, the summer birds have left for warmer places. The natural world is preparing for the cold winter months ahead, and so too must we.

I love being outside, and being outside with children is even better. The outdoors have so much to offer, and the opportunities for learning and personal growth seem endless. The first day I took a class of students outside for an extended period of time it was magical. Many things changed for me that day, and I knew that I had to continue this new routine, even during the winter. When I think back to the first cold day outside with my students, I remember how hard it was, and how I didn't think I'd be able to keep doing it. I also remember how much fun we had that winter, because I decided to keep taking my first-grade class outside. That year I had thirteen energetic students, and being outside allowed everyone in the group to get what we needed. The kids could move in any way they wanted, and they could be loud, or they could find a quiet spot out of the way to talk and play quietly with a friend. I started teaching because I wanted to support children as they grew; I wanted to help them learn to read and discover interesting things about the world around them. I found that taking my classes outside was exactly what I needed to be the best teacher I could be. Eventually, I was able to be more present, experience more joy, and build even stronger relationships with my students.

I remember one winter day in particular: It was early December, and it did not feel like a very cold morning. It was sunny, but breezy, and there was frost on the ground. We hiked the long way out to our basecamp and so our bodies were warm. A child—who earlier had refused to put on snow pants—sat on a log, then took off their mittens, hat, and, finally, their jacket, proclaiming, "I'm hot!" The other children followed suit and also started to take off their layers, and I started to worry about the rest of our time outside.

I was worried that the children would get cold and, further, that our outing would take a turn for the worse. While the children were too hot now, they were going to be cooling down in our next activity. We always start our time in the forest in a circle with a song, story, discussion, and snack. We don't spend a long time in circle, but it is long enough that our bodies start to cool down. On this day, we added in an already cool morning, some shade, and a breeze, and we had a recipe for very cold kids.

Some children who had taken their coats and hats off soon realized they were feeling cold and bundled back up. Others just sat there, feeling discomfort but unable to make a decision about what to do, seemingly paralyzed by the cold. It took some time and some support but eventually everyone had put their coats, hats, and mittens back on. We then moved our bodies to bring the warmth back into our fingers and toes, and we were able to enjoy the rest of our time outside that morning.

This experience became an important lesson for all of us. As the adult, their teacher, I knew that I needed to set boundaries around taking off layers. I also knew I needed to help the class brainstorm what to do when they start to feel too warm or too cold. I decided the best way to do this was to involve the class in the process, in much the same way that I did in setting the classroom agreements.

In our first meeting we produced lists of what to do when we are feeling cold. The students' ideas included running, jumping, adding another layer, and having something warm to drink. At our next meeting, we wrote a list of what to do when we are feeling too warm and uncomfortable. Writing these lists required more support from me as the adult because "taking everything off" was not an option. I wanted them to realize that removing layers was an option but needed them to understand that

removing too many layers wasn't the best idea, nor was it safe. We ended up with a list that included ideas like choosing to unzip jackets, take off hats and mittens, and waiting a little to see how our body felt before doing anything else.

In a different meeting we talked about paying attention to the weather to determine what we needed to wear outside that day. We learned through the winter that the sun made it feel warmer and wind made it feel colder. If it was sunny and windy, it would probably be cold; a cloudy day with no wind could feel warm. There was still a lot of learning to do in regard to being outside in the winter for extended periods of time. What we did learn together was that being properly dressed, having a plan in place for what to do if someone is too cold or too hot, and making close observations of the weather all made a huge difference in our experience outdoors.

Winter in the Northeast brings many opportunities for us to learn, grow, and have fun. The change of seasons, in particular, can bring about great opportunities to talk with students about dressing for ECO. It is almost as if the shift from late fall to winter is Mother Nature's way of saying, "Okay my children, time to bundle up! Colder days are coming."

Taking children outdoors for extended periods of time in cold weather requires preparation and dedication. It is hard! There is a lot to think about: they might get wet, they will most certainly get cold, maybe really cold. What will we do? But beyond those concerns is this: they might

Kelsey LaPerle

get to experience the tiredness that comes only from trudging through deep snow, and they might get to catch beautiful snowflakes on their mittens. They might get to see and feel the way the ground squishes when it is really wet and then how hard and crunchy that same ground feels when it freezes. They might get to experience the sound of rain falling on the leaves, or their hoods, or the metal roof of a nearby building.

In my experience, the benefits of being outside in all types of weather are worth all the perceived and real hassles of getting kids dressed and out the door. Here are just a few benefits associated with teaching outdoors for both teachers and students:

While at ECO it started pouring, one child commented, "It's raining!" Without a pause, every child put up their hood and continued listening to the story without complaint. At the end of our rainy morning, I was walking back to school holding hands with a little girl who was making up a song that went something like this, "I love nature, I love nature so much! We got to be in nature. I love nature so much! Nature is so perfect! Nature brings me joy! My teacher brought us to nature and that is why I love her!"

—*Erin Gale, teacher, Morristown Elementary School*

Amy Butler

- Sing a song about the weather. Make up skits about weather. Celebrate the task accomplished and the weather of the day. I once worked with a teacher who brought out a speaker and played "Singing in the Rain" by Gene Kelley. Needless to say, we had a dance party!
- Staying warm is crucial, and it can also be a learned skill for children that provides a way to tune into their bodies and think about how they can care for themselves. One helpful tip is to have movements depicted on laminated cards that students can pull from a bag if they need ideas for warming up. Examples are jumping up and down like a frog ten or thirty times, flapping their arms like a hummingbird for one minute, or lumbering like a bear across the forest and back. Now, bring a friend!
- Go on a weather protection scavenger hunt. Look for places where you can be sheltered from wind and rain. Where are the dry and sunny areas on a winter day in your outdoor classroom? Do you think animals might spend time in those places too? Why?

We Can Use These Materials Outside

- Bring thermometers, rain gauges, and snow measuring instruments. Have a large thermometer handy for reading indoors and then bring it outdoors and see how the temperature changes.
- Have samples of different types of material that our clothes are made of. Experiment with laying them in snow, misting them with water, or soaking them in a puddle. What happens to wool or cotton when it gets wet? How does a raincoat compare to a windbreaker to keep us dry?
- Bring a scale to weigh wet sample materials, such as wet forest debris compared to dry forest debris.
- A piece of string with a feather tied to the end is a simple way to read wind direction.
- Don't forget a thermos of tea and everyone's teacups!

When We Get Back Inside

- Recording weather is an opportunity to practice phenology over the course of the entire school year. Phenology is the study of seasonal natural phenomena that includes plants, animals, and weather. Get into the habit of recording the weather as a class.
- Interview the local meteorologist. Have children come up with questions and send these ahead of time. Make it a seasonal interview related to the weather changes happening each season.

- Become nature pen pals with another class that lives in an entirely different region from you. What could a class of Vermont students learn from a class of students in Arizona?
- There are so many STEM lessons related to weather. A simple search on the internet will bring you to many activities to use inside the classroom.

Books about Weather and Dressing

Do I Have to Wear a Coat?, by Rachel Isadora

Warm Winter, by Feridun Oral

A Day So Gray, by Marie Lamba

When This World Was New, by Danilo Figueredo

Lesson 18. Becoming Bears

Narrative by Ken Benton

The trees were bare, and temperatures were beginning to drop. The kindergartners at Moretown Elementary School turned their focus to how animals prepare for winter. In their classroom, they had just examined the differences between fiction and nonfiction by sorting photos of real bears and the pretend bears they had seen in cartoons or read about in stories. With images of lumbering bears fresh in our minds, we decided that the best way to learn how bears prepare for winter would be to become bears ourselves, utilizing the power of imaginative play.

Nick Neddo

The morning of our ECO session, the classroom teacher led a discussion on the basic things that animals need to survive, which are the four elements that make up a habitat: food, water, shelter, and space. The class was told they would be going on an adventure in the afternoon and that they would have to pack a bag with food and water. While the children were outside for recess, bedsheets were draped across some of the desks, and pillows were laid out on the floor underneath, setting up a makeshift den. Gathering the students under the sheets upon their return, we had the students close their eyes and imagine thick black fur sprouting out all over their bodies, fingernails growing long and sharp, and noses becoming more powerful. They were now a class of kindergarten bears. Crowded into our small cave, we revealed that they had grown too large for the shelter, and the time had come for them to find dens of their own. We packed up our food and water and set off from the school and up the hill, lumbering into the woods in search of fresh dens for a long winter's nap.

Just like young bears venturing out on their own for the first time, we were exploring a completely new territory, a section of forest where we had never been before. Once we found a suitable area to set up our base, we left our bags in a pile at the base of a tree and began crawling around the forest floor. The little bears searched under boulders, inside hollow logs, and in deep dark caves for a shelter that would keep them dry if it rained, block the howling of the cold North Wind, and keep the snow from falling on their slumbering bodies. Every potential den site was examined with their big bear noses to be sure there wasn't already a large hibernating occupant inside. Each den was thoroughly tested. How many bears could it hold? Was it comfortable enough to sleep in? If it wasn't, leaves were gathered and piled up to make a soft bed. One little bear actually fell asleep, curled up on their pile of leaves.

Once claim was laid to a den, it was guarded with the utmost ferocity. Even the meekest, mildest student in

"Are black bears warm?" questions a student.

"Do you mean right now; are they warm today?" I ask.

"No, I mean are they warm? Are bears warm?"

I take a moment to think, then reply, "Yes, I think a bear is warm. Their fur is thick and cozy."

The student smiles lazily, snuggles down into their neck warmer, and lifts their shoulders. "Mmm, that feels good for a bear and me!"

the classroom had transformed into a mother bear with the most terrifying roar of them all. Anyone daring to enter a ten-yard radius of her den was quickly turned away with a menacing growl and gnashing of teeth.

Ken Benton

At the end of the day, we walked back down the hill and into the classroom. The spell had been lifted, and even the cantankerous mama bear had transformed back into a human child. Before leaving for the day, we made two lists: The first was a list of the needs of bears, and the second was the needs of people. As the lists formed, we began to notice they were not that different from each other. The next morning as review, each of the students drew a picture of their bear den, including the food and water a bear would use to survive, and compared it with a drawing of their own home and what they (the human students) eat and drink.

Through creative role play, students connected to the land in new ways. By taking on the persona of a young bear venturing off into new territory for the first time, they were able to engage with material in a way that would not have been otherwise possible. Though they were children once more, they would always remember the afternoon they became bears.

Becoming Bears

Grade Level: K–First

Objectives

- Students will understand that all living organisms have basic needs that must be met for survival.
- Students will understand how animals use their external parts to help them survive, grow, and meet their needs.

We Are Exploring These Academic Standards

NGSS.K-ESS2-2. Construct an argument supported by evidence for how plants and animals (including humans) can change the environment to meet their needs.

NGSS.K-ESS3-1. Use a model to represent the relationship between the needs of different plants and animals (including humans) and the places they live.

NGSS.1-LS1-2. Read texts and use media to determine patterns in behavior of parents and offspring that help offspring survive.

NGSS.K-2-ETS1-2. Develop a simple sketch, drawing, or physical model to illustrate how the shape of an object helps it function as needed to solve a given problem.

What

This lesson is inspired by watching children consistently pretend to be animals in their play schemas. We see young students, and at times older students, leaping like cheetahs, howling like wolves, and roaring like lions and dinosaurs. Children get closer to understanding nonhumans by becoming the animals they admire and wonder about. David Sobel, in his book *Beyond Ecophobia*, speaks to the importance of connecting to local species by creating rich learning opportunities and allowing imaginative play to unfold. Imitation of animals incorporates physical and mental planning to act out behaviors and movements. We are encouraging the students to consider the bear's perspective as it gets ready for winter. By learning about bears and their adaptations for winter, children can connect their needs to the needs of other living things. Hopefully, this will encourage students to care for themselves and dress properly for outdoor learning in winter!

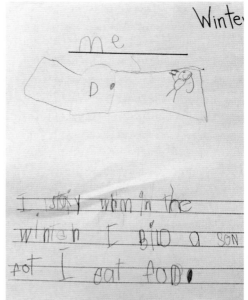

Amy Butler

How

We can begin this lesson by discussing with students what our needs are to survive (food, water, shelter). Where do we get our food and water, and what are our shelters? How are our shelters built? What are the materials used? How do we stay warm in our shelters during the winter? What materials are used to keep us warm?

Next, discuss the needs of bears. What sort of shelter do they live in? How is it built and kept warm? Why do animals need shelter? What makes a good shelter site? Exploring the difference between fiction and nonfiction bears in literature can help students visualize and verbalize their adaptations needed to survive as a real bear. Are we black bears, brown bears (grizzly bears), or polar bears? In preparation for turning into bears and looking for dens, students can make a plan for what their den will look like, where it will be located, and how it will meet their needs for winter.

Now you may begin a role-play that will carry through the rest of the lesson.

From here on out, they will pretend to be young bears. An adult can pretend to be the mother or father bear and try to gather everyone underneath a tarp or blanket shelter set up before the lesson (the shelter should be made small enough to just barely fit everyone inside when squeezed together). Explain that there is no longer enough room inside the bear family den for everyone and that the time has come for the young bears to find shelters of their own for the winter. Students will pack their backpacks with water bottles and snacks in preparation for the hike to search for dens.

Spend the remainder of the time in search of good den sites. Explore your local park, town forest, school grounds, or public wild spaces for boulders, hollow logs, ledges, and the like, and discuss each potential den as you find it. Would it keep out the rain and snow? Would it stay warm inside? Is there enough space? Students should try to fit inside each den site to try it out. This is a chance to venture into spaces that have not yet been explored. It is the teacher's responsibility to perform a site assessment of any new area you are visiting.

Encourage imaginative play during the search through the use of all senses, especially the use of "bear noses." After extensive exploration, if a suitable den site or sites are found, allow students to use leaves or sticks to make them more comfortable for their winter sleep. Whenever most convenient, take time to sit together and eat snack in the woods and share stories about how you found your den and what it will be like sleeping all winter in it!

We Can Use These Materials Outside

- If you do not have access to a wild space that has possibilities for mimicking a bear den, bring sheets, blankets, or cardboard to construct bear dens.
- Bring laminated pictures of bears and field guides with bears in them. A visual reminder of what bears look like, and their habitats, will heighten imaginary play.

When We Get Back Inside

- Make simple bear masks and bear paws from cardboard or paper. Dress as a bear next time you go out to visit your bear dens.
- Create a bear habitat and den diorama. Include all the things needed for a bear to survive the winter.
- Check out the North American Bear Center (bear.org) and the Wildlife Research Institute (bearstudy.org)

located in Ely, Minnesota. There are live bear den cams and lots of bear study videos.

- Learn about animals that are true hibernators in your region. Who is sleeping, and who is active in the winter?
- Contact your local fish and wildlife department and find out if they offer educational materials on fur bearing mammals in your state. Some states have kits for teachers that include pelts, skulls, track sets, scat sets, and field guides.

Books About Bears and Winter Hibernation

Time to Sleep, by Denise Fleming

Bear and Wolf, by Daniel Salmieri

Winter Sleep: A Hibernation Story, by Alex Morss and Cinyee Chiu

Animals Hibernating: How Animals Survive Extreme Conditions, by Pamela Hickman, illustrated by Pat Stephens

Bear-ly There, by Rebeka Raye

Lesson 19. Who Goes There? Animal Tracks in Winter

Narrative by Amy Butler

ANIMALS LIVE IN THE FOREST BEHIND OUR SCHOOL.

This was the only statement written on the board inside the classroom one morning. The children, who were dressed in wintery layers for ECO, left their coats, hats, and mittens off and settled into our morning circle. We brought the attention of this group of third graders to the statement written on the board. They all looked at the adults with an expression of, "Ya? And so what?" One student even replied, "We know that already!"

We explained to the students that we would be flipping things around that day. Instead of an adult asking a question about what students know about animals in the forest behind the school, the students were going to come up with their own questions based on the statement we wrote on the board. This was a new approach for this class, and

Nick Neddo

you could almost hear the gears in their minds screeching to a halt and trying to recalibrate. This approach is inspired by the Question Formulation Technique (QFT; this is trademarked), developed by the Right Question Institute. The QFT is designed to help students improve their skills around asking questions and to feel empowered to ask more.

We had already seen these students showing interest in tracking and trailing animals in the snow, and with a refresh of light snow overnight, we wanted to tap into this in the outdoor classroom. Since winter was here, and snow would be consistently covering the ground, it was the perfect time for tracking. To set them up for this outdoor activity, we had the students start to generate questions about the statement during our morning meeting. It was a little slow at the beginning, but by following the QFT

Amy Butler

rules we were able to write a robust list of questions.

Students asked: Where do they sleep? How do they survive? Will we see their tracks? How do they catch their prey? What games do they play? What do they do? What are their predators? Where do they go during the day? Where do they find water? How do they find food?

There were bound to be many signs of animals moving about, and our hope was that today we could harness some of the student-driven inquiry and learn how

Question formulation technique rules

- Ask as many questions as you can.
- Do not stop to discuss, judge, or answer the questions.
- Write down every question exactly as it is stated.
- Change any statement into a question.

to do some tracking without stepping on the evidence. This can be a challenging skill for students to learn, and this group of enthusiastic third graders needed some practice! Noticing that many of the questions involved animals moving and searching to meet their basic needs, we decided it would be a good idea to introduce an activity on how to follow an animal's tracks.

With our questions in our pockets, we headed to the forest and gathered in our meeting spot for directions. Students had already noticed fresh tracks in the snow and were eager to get exploring. We usually play a game in the field before entering the forest, and today we decided to use this activity in place of our game, as it was active with a certain level of focus. Students would be moving about and cooperating with a partner to lay a trail of clues for another pair of classmates to find. In our meeting area we reviewed what tracks we had noticed last week and how we had followed them. We asked what was hard about following tracks and what worked.

"I found a deer trail, but everyone kept stepping on it and I lost it," said one child.

"Yeah! Me too!" said another. "I saw a bunch of tiny tracks all around trees and they looped around and around. I wanted to follow them!"

We explained to the students that they would be working with a partner to lay their own trail for another group to follow. They could use sticks, arrange leaf patterns, or use some of the materials in the bag we gave them. Inside small cloth bags were a collection of bright green pieces of yarn, brown craft feathers, and squares of cloth. They would need to make a total of twenty clues for others to follow. Clues should be close enough to see from one clue to the next and should stretch into the forest without overlapping another group's trail of clues. We then had all the students close their eyes, and we laid a trail of clues right through the center of our circle, heading north into the forest. Even from where students were sitting, they could see all five clues we laid out. Our clues were sticks set in an X pattern and also sticks in groups of three, laid out all pointing the same way. This was a way for us to give an example of what it might look like and to demonstrate how to move along the trail, pause and look, and not step on the clues. We emphasized that the clues should be visible but not super obvious. We wanted the students to practice slowing down and seeing the winter forest floor as a blank canvas and also to notice

its shadows, the different slopes and elevations, and the blown-down treetop debris.

We hoped this game would not only be fun and exciting, as if laying out a treasure map, but also give the students an experience in using pattern recognition and subitizing in the act of tracking. We experience our surroundings through our senses, and so that day we focused on sight and how that informs our internal understanding of the world around us. And the students had produced a whole list of questions about animals living in the forest back inside the classroom! How can tracking, learning how to slow down, and following a story help us answer those questions this winter?

Students scattered and got to work laying clues, adjusting distances, and keeping sneaky classmates from peeking at their work. Soon partners were ready to trade trails and test out each other's work. The voices in the forest changed as they became seekers and trackers. There were long pauses of silence and then bursts of exclamation as clues were found and followed. Creativity was applied at the end of the trails with elaborate nature mandalas and words spelled with sticks, like "YOU DID IT!"

Whenever a lesson goes over well with students, we see it repeated during free time. The students flowed into our next part of the morning, with the fire tenders collecting mouse tails and chickadee sticks for a fire, and some of the students continued to lay trails of clues for others. Other students decided to follow animal tracks, and a few grabbed their journals to bring along. I watched students approach the forest a bit differently from the week prior; I noticed they were pausing, stepping slowly, and looking out ahead and all around for animal clues and answers to their own questions.

We can easily spend a whole morning trailing animals and telling stories from the tracks left behind. Tracking becomes part of our winter routine. We are building inquiry skills and, more importantly, quieting the mind as we consider the other more than humans we share this landscape with.

Who Goes There? Animal Tracks in Winter

Grade Level: K–Third

Objectives

- Students will understand that animals leave behind clues that help us understand their behaviors.
- Students will be able to follow a trail of clues in sequence, from beginning to end.
- Students will learn how to recognize disturbances in the forest made by humans and animals.

We Are Exploring These Academic Standards

NGSS.K-LS1-1. Use observations to describe patterns of what plants and animals (including humans) need to survive.

NGSS.2-LS4-1. Make observations of plants and animals to compare the diversity of life in different habitats.

NGSS.3-LS3-1. Similarities and differences in patterns can be used to sort and classify natural phenomena.

What

After pretending to be bears and squirrels, building rabbitats, and adopting nature names in the previous lessons, students will be ready to explore the lives of the unseen animals that live among us. On our outings we may be lucky enough to see a squirrel scamper through the branches or a bird alight on a branch near us. But what about the resident fox, deer, snake, or raccoon?

Tracking animals taps into children's curiosity and the drive to hunt for clues to mysteries. We know science investigations begin with a question and learning to follow animal tracks begins with many questions. Who lives in the forest, field, or park? What are they doing and when do they move about? Where are they going and why?

This lesson will springboard students into the world of tracking animals. A good introduction to any new skill is to make it fun and accessible. This lesson helps students to learn how to travel as a group along a trail without disrupting the tracks, or the "story," as we like to call it. Following an animal trail can be like reading a story and we don't want to step on it!

Amy Butler

How

In this lesson students are making a trail of clues to mimic animal tracks and other animal signs. Students will work in small groups or pairs to lay out a "story" for other students to follow. Discuss what signs of animal activity we may find while tracking. Tracking is not limited to just footprints! Animals that have been close by may leave fur, scat, chew or rub marks on trees, scratches and scuffs in the dirt, beds or lays, burrows, cavities in trees, midden piles, feathers, fallen nests, and even bones or a skull!

Explain that students will use some natural items (sticks, leaves, or stones) to make their own animal tracks and signs along a trail. Sticks can be made into arrows or letters, so they stand out along the trail. Stones may be stacked, and leaves can be arranged in a pattern. These are natural and subtle clues that students need to slow down and search for. This requires that students work together to not step on clues in order to sequence

them and see what direction they are headed.

Students can also use unnatural items (string, cloth, colored feathers) to help their peers stay on the trail of clues. This is especially helpful for younger students!

Have students plan which direction they will lay their trail. How far apart should the clues be? Are you able to see them? What is the destination of their trail?

Have the students choose where the trail begins and where it ends. Be aware that some trails could cross over others depending on the amount of space you have. This can easily be done in a school yard or along a sports field.

Once students have laid out all the signs, have them go back over the trail slowly to see if they can follow it themselves! They should have a clear start and end to their trail of clues. Afterward, find another group that has finished their trail and switch. Ask the students questions, such as: What clues did they leave behind for you to follow? Was it an easy trail to follow or a challenging one? Why?

What happens when we follow animal tracks?

- Students connect directly to the animal species close by
- Focus and concentration heighten
- Students visualize and then verbalize their discoveries
- Pattern recognition develops
- Observation leads to critical thinking and inquiry
- Students lead, gather evidence, and solve problems

Make sure students erase their trails and collect all non-nature items when they are done.

We Can Use These Materials Outside

- Pieces of colored yarn, scraps of fabric, or craft feathers
- Field guides that show animal tracks and signs animals leave behind (these can be used as reference when laying out the clues)
- A measuring tape to distance the clues with consistency

When We Get Back Inside

- Have students map in their journal the trail they laid and the trail they followed.
- Use guidebooks to study how animals leave trails behind in the forest, and what those trails can tell us about that animal's behavior. Students can use their trail to create an adventure story, written or oral, focusing on sequencing vocabulary.

- Play a game of animal charades: Students can act out animal behaviors, and classmates can guess what sign may be left behind after a specific behavior is displayed. Is the animal digging for food, making a bed for the night, or building a nest?
- Where are other places to follow animal trails? Are there areas with sand or muddy, soft ground in your neighborhood? Animal tracks can be found near water, along trails that humans walk, and in gardens, too.

Books about Tracking and Animals in Winter

Animals in Winter, by Henrietta Bancroft and Richard G. Van Gelder

Winter is Coming, by Tony Johnston

In the Snow: Who's Been Here? by Lindsay Barrett George

Snow Secrets, *by* Lynn Levine

A Wolf Called Wander, by Rosanne Parry

Lesson 20. Finding the Mystery

Narrative by Dave Muska

Today, we are focusing on animal tracks with the kindergarten class. It's late November and the mountain tops are frosted with a light snow, but in the valleys, where we are, the ground is still bare. It is the rut, also known as white-tailed deer mating season.

I pass a deer antler around our circle as I share the story of finding this treasure, of following a deer trail along a quiet and still river in late winter, through tall grasses and brush. Inspired by the story, we spread out to explore our forest classroom, the children excitedly looking for tracks and signs of white-tailed deer. Their eyes are focused on the forest floor covered in this season's release of leaves—birch, maple, and beech. The children naturally gather into groups of two or three, just as ancestral hunters have done for eons. Their voices carry a mixture of excitement, wonder, and embodied freedom with the task ahead.

Nick Neddo

Finding animal tracks on the forest floor in the leaf litter can sometimes be a challenge, so I locate a deer track and call out, "Wow! Check out what I found!"

This is a great way to naturally call the class together. They run toward me, and I pause them as they arrive just before they reach the track.

"What shape is that?" I ask.

"It looks like a heart . . . or a triangle . . . it's a deer track!"

The other students clamber to see it, not believing that an animal's track could be visible in the fallen leaves without snow. "Can you see it?" I ask motioning to the ground. "See what?" says a student. "It's a track. I'll show you how to find it," I say.

I ask the students to spread out in a line along the deer trail. They can't quite see the tracks or trail yet.

"Look at the ground in front of you and look for the animal tracks," I explain. The students start looking on the ground

These field guides will help teachers and students understand the basic movements of the four gaits:

- *Mammal Tracks & Sign: A Guide to North American Species*, by Mark Elbroch
- *Tracking and the Art of Seeing: How to Read Animal Tracks and Sign*, by Paul Rezendes
- *Peterson Field Guide to Animal Tracks*, 3rd ed., by Olaus J. Murie, Mark Elbroch, and Roger Tory Peterson
- *Mammal Tracks and Scat: Life-Size Tracking Guide*, by Lynn Levine and Martha Mitchell

planning their next move. They are autonomous. Have them lope like a bobcat.

There are of course many more benefits within this lesson, like staying warm, tapping into dramatic play, building a foundation of tracking skills, and growing an appreciation of the animals that live close by.

How

Start with a conversation about how students think animals move. Have they watched a dog or cat move? What about a squirrel dashing across the road or nimbly moving along a power line? What are the many ways that humans move? How do animals move in order to survive?

From here, you can introduce the four different ways in which animals might move: walking, loping, bounding, or galloping. These are called gaits. Gaits are the movement patterns animals use at different times. By classifying animal movements into gaits, it helps us, the

trackers, learn about what the animal is doing and who the animal is. We could focus on the single track to try determining who the animal is, and we also have an opportunity to follow the trail and read the story of how the animal moves across the landscape. Where is it going, and how is it feeling? These are questions that go beyond asking, "Who is this animal?" to more in-depth investigation, such as, "What else can I learn about the animal by following its movements?"

To start, know that these movements are simple to act out, and students and adults have the freedom to determine what they look like. Make predictions, move your body imitating the gait, make adjustments, and ask questions. Teachers can even use the QFT mentioned in lesson 19, Who Goes There? A statement such as "Animals move through deep snow" can be a launching-off place from which to discuss animal gaits and generate questions we have about how animals move. Discuss with the students the animal's shape and size. How might that animal move? What characteristic does the animal have that helps it travel through snow?

The goal here is to practice moving like the animals that live near us in order to better understand them, connect with them, and further develop empathy for all animals. None of us will have the time to become expert trackers over the course of this unit, but we can "put the quest back in question and the search back in research," as outdoorsman, author, and animal tracker Tom Brown Jr. has wisely said.

We Can Use These Materials Outside

- Make laminated mammal cards of the common species in your region. Bring these into the outdoor learning space and always have them on hand in the classroom. These are excellent prompts for storytelling, movement games involving gaits, and practicing our inquiry skills. Consider that young students may already know many corporate logos—but can they identify twenty local species of animals?

- Examples of the four gaits can be traced onto yoga mats or the classroom floor. Start by naming the four animal gaits (walking, loping, bounding, and galloping). Look up an example of each gait and, using a black permanent marker, draw the track and gait on the mat or use masking or duct tape on the classroom floor.

- Students can practice the gaits in fresh snow, but be careful not to step on other students' tracks! Then, using a tape measure, students can measure their stride and the width and length of their track.

- Bring a camera on the tracking exploration to capture the experience. Use these pictures to create a story, either written or oral. This can also be a springboard for researching local animals and creating a classroom guidebook.

When We Get Back Inside

- Watch videos of a common animal in your region, such as squirrels, coyotes, deer, or cottontail rabbits or hares.

- Match students' nature names, from lesson 5, to this lesson. Choose two students each week to focus on and highlight their animals' movements and actions in the winter. Do they stay active, hibernate, migrate, or go into torpor during the winter?

- Use purchased tracking stamps or draw tracks freehand to create a scene, and then tell a story about the scene. You can find accurate wildlife track stamps at Acorn Naturalists (acorn naturalists.com) and Nature-Watch (nature-watch.com).

- Students can make their own tracks by painting their feet and hands. Then they pick a gait of a four-legged mammal and move across a large strip of paper using their chosen gait. Once the paint has dried, they can measure the stride, length, and width of their tracks. This involves a lot of supervision, and is a bit messy, but your students will love it!

Books About Finding the Mystery

We All Play, by Julie Flett

Tracks in the Snow, by Wong Herbert Yee

Over and Under the Snow, by Kate Messner and Christopher Silas Neal

Big Tracks, Little Tracks: Following Animal Prints, by Millicent E. Selsam and Marlene Hill Donnelly

How to Be a Nature Detective, by Millicent E. Selsam and Marlene Hill Donnelly

Lesson 21. The Tale of the Three Snowballs

Narrative by Amy Butler

March. We are heading into another shoulder season here in the Northeast. The ground manically freezes and thaws. We put rain pants on over snow pants in a desperate attempt to stay warm and dry. Our winter patience is growing thin, and managing all the winter gear now covered

We had explored snow and its properties all winter. ALL winter because that is what it feels like here in Vermont. Back in late November we started having conversations about snow. We called it "What We Know About Snow." Kindergartners were practicing having

Nick Neddo

in mud is tiresome. The kids want bare feet, no jackets, and mitten-free hands. We still have quite a way to go before that. Thank goodness for the longer days and warming sun, the return of the red-winged blackbirds, and the running of maple sap. Drip, drip, drip. We can hear the snow melting! Even though the calendar officially declares it's spring, the snow melts and then piles back up again in a late winter storm. Children are wondering why the snow sticks around. Will it ever melt? And will it be here tomorrow?

a discussion without raising hands and waiting for a teacher to call on them, but rather listening, anticipating, and taking turns in small groups.

These conversations continued because we had weekly experiences in snow: hiking to our outdoor classroom, building with snow, watching it melt and refreeze near our fire area, and tracking the resident squirrels and red fox in the snow. Students can build their under-standing of properties of matter, such as snow, by simply being in it every week. When we practice inquiry, encourage

Kelsey LaPerle

Kindergarten students talking about what they know about snow

When sun shines on snow it sparkles.

If the sun is very hot it can melt the snow.

One time I saw two snowflakes that were a circle and three lines that were poking in, and it was the same thing.

Once in preschool we used these things to dig into the snow and we couldn't because it was too hard, so I crawled. I wonder what made the snow so hard and icy?

The sun melted it and then it turned cold and then it turned icy.

Sometimes the snow is sticky in one part and hard or slippery on another. . . . I don't know how I found out how it was sticky, but I put both of my hands on the snow, and it got sticky.

I didn't know that snow was sticky . . . I kind of believe it and I kind of don't. I'm in the middle. I don't think sticky snow is real.

How do you make ice into sledding snow, like at recess when it's really slippery?

I didn't know that snow might be sticky enough to stick onto us or hold on to. . . . How does snow get sticky?

I wonder if snow is really, really sticky and hard enough and you sit on it if it would break or not?

How many snowflakes does it take to make a snowball? How do you build a snowflake out of snow?

I have never made a snowman before . . .

Well, I don't know how to make snow dog.

Is snow really melting diamonds?

Unit 5
What Does Spring Bring?

The beautiful spring came,
and when Nature resumes her loveliness,
the human soul is apt to revive also.
—*Harriet Ann Jacobs*

It is spring. Everything is waking up, and there is a collective sigh of relief here in the Northeast. The level of comfort and confidence teachers see in their students is much different than in September. At this point in the year, they have developed a familiarity with the landscape with which they have been frequently engaging. Insects and spring ephemerals are gently cared for. Students' core strength, stamina, and decisions around risk-taking are stellar. Their relationships are flourishing too. Students who don't normally engage with each other in the classroom are working together to engineer solutions to problems and create new games and play schemas. There's little concern about dirty hands or clothes, and toes and fingertips are warm. Unit 5 brings us to the end of the school year and to the beginning of more discoveries as the earth softens and expands once again.

Lesson 22. Between a Stump and a Wet Place: The Scientists' Coverboard

Narrative by Pete Kerby-Miller

"WOLMS!"

All the preschoolers rushed to Micah's overturned stump. Our circle and snack routine was quickly forgotten amid the excitement of an animal visitor. The beginning to our day in the forest just took a new direction.

I glanced at the pizza-box-sized plywood sheets I had planned to introduce during our stump circle meeting and began to draft a new flow for the lesson, while covering the few steps across the circle. Students surrounded the tipped stump, crowded shoulder to shoulder on their knees, and circled around the overturned stump to get a better look. The group bubbled with curiosity.

"There's two of them!"

"I found them!"

"I can't see!"

"They're so shiny!"

"I can't see!"

"Whoa cool! Earthworms!" I exclaimed upon reaching the crowd. I then dropped my voice to a near whisper and asked, "Why do you think they are right here? Under Micha's stump?"

Those who had been circling for a better view found places between their kneeling classmates. As my questions hung in the air, the eager rush settled and gave way to contemplation. Shoulders dropped and a few eyes were cast up in thought. Children whispered replies.

"They're eating the dirt."

"This is their home."

I noticed that little hands, which had been pinned under little knees as a way to resist the temptation to grab, were now returning to laps. At the start of the school year, we had been learning about the living things that had their homes in our outdoor classroom. Preschoolers had considered what it would be like to be a small creature, like a worm, and to be grabbed and lifted up away from your home. How would it feel to have many loud and big voices shouting over you? With weekly practice and much reflection, these young learners were able to apply the ways of "A Gentle Giant" (see unit 1, lesson 2, Catching Empathy). Since Micah was the one who found the worms, they were also the one responsible for the care of the worms. With so many children around these few worms, Micah would be the only child, at this moment, to touch the worms. Since this was established early on as a way to be with and care for the living creatures around us, it has become a part of our outdoor classroom culture. The children had become accustomed to pausing with quiet voices and waiting with patience to know more.

Together, this bundle of curious preschoolers talked about what makes a worm home, like darkness, moisture, and dead leaves to eat. The underside of Micha's seat had everything worms needed. With their new knowledge, preschoolers were ready to set off worm-searching. First, we restored the stump-bottom habitat our first worm neighbors had enjoyed before the discovery. Lifting a stump wasn't quite as easy as tipping one over, so a few determined preschoolers teamed up to help set it upright. With a satisfied pat to the stump, the last student turned away and joined the exploration in the forest classroom for worm habitats.

Preschoolers set out into the forest classroom looking for worm habitats. "Too dry," said a student after peeking under wood in the woodshed. Although, the underside of the cable spool table was home to plentiful worms, as was beneath a branch marking a path.

"Whoa! Whoa, look at this!" I followed the sound of an excited voice to where several children were staring agape at a treasure. Nestled under a small log and next to a small twig, was a cluster of two dozen pearly orbs, each one not even half the size of a pea.

I knew this was a cluster of slug eggs, but the identity of an interesting organism is not often something I'm willing to divulge so quickly. Often, students'

curiosity stops once they have an answer. If I were to share the name of the sticky bundle, I might preclude observations of structure and consistency, questions of origin or purpose, and a myriad of imaginative stories. Many creatures are inherently interesting, and these pearlescent eggs were certainly doing the work of inspiring a sense of wonder on their own. For animals so vastly different from ourselves though, a bit of information can help build empathy and spark imagination. I soon sensed it was time to move this observation deeper.

I shared that we were looking at slug eggs, and those students dashed off excitedly to bring their friends over to their find. I stole back to the meeting area to retrieve those plywood sheets I had brought out to the forest. When I returned, the class had gathered around the slug eggs.

"I lifted this log and there they were! Slug eggs!" a child exclaimed to their peers.

Kelsey LaPerle

The class was enthralled by the egg cluster, so I decided to complement their inquiry while shifting focus back to habitat.

"Why do you think a slug laid their eggs right here?"

We shared what we thought slugs might need, such as food and a place to hide from the sun. Someone pointed out that this log has an excellent view of the soccer field. Very important indeed!

"Some logs can be hard to lift, so I brought these boards to make our own slug and worm habitats. We can even put our scientist hats on to experiment with what materials we put underneath. What do you think will make a perfect slug home?"

Groups of preschoolers found places for each of the boards and volunteered materials for each. Delicious dead leaves were popular, though one group set theirs up on a grassy surface, because grass is Liza's goat's favorite food. The student who found the slug eggs placed a small twig under each, hoping to find more pearly treasures next week.

We then returned to the stump circle and the snack that we had abandoned in the worm-induced excitement. As we nibbled, children shared what they thought might happen under each coverboard. Would worms like the grass? Would a millipede move into the bark scraps like the one found by the fort-building area? With such burning questions awaiting our return, we settled on a new routine for our weekly forest class time: first we'd check for visitors under our coverboards, then we'd have snack and share what we found. We all left the forest that day full of excited anticipation to return.

The day didn't follow the schedule I had written out, and while the rush to see the "wolms" disrupted our usual routine, I could never replicate such an engaging call to investigate that which wriggles and crawls beneath our feet. Be they students or slug eggs, we always have co-teachers in the forest classroom.

Between a Stump and a Wet Place: The Scientist's Coverboards

Grade Level: Pre-K–Third

Objectives

- Students will understand that humans share the environment with animals and plants.
- Students will construct concepts about the characteristics of living organisms and ecosystems through exploration and investigations.
- Students will monitor coverboards on a weekly basis by tallying and counting living creatures.

We Are Exploring These Academic Standards

NGSS.K-LS1-1. Use observations to describe patterns of what plants and animals (including humans) need to survive.

NGSS.1-LS1-1. Use materials to design a solution to a human problem by mimicking how plants and/or animals use their external parts to help them survive, grow, and meet their needs.

NGSS.2-LS4-1. Make observations of plants and animals to compare the diversity of life in different habitats.

NGSS.3-LS4-3. Construct an argument with evidence that in a particular habitat some organisms can survive well, some survive less well, and some cannot survive at all.

What

Have you ever wondered what goes on in the world beneath our feet, or what takes place under a rotting log? There are millions of tiny creatures living in these spaces, breaking things down and fertilizing the plant life. And we know that by simply flipping over a log, whether it's in a manicured yard or a forested area, we can see worms, insects, spiders, and small mammals.

Scientists use different ways to study the world, and a coverboard is a tool often used by herpetologists to attract reptiles and amphibians. A coverboard is exactly what it sounds like: a wooden board or piece of sheet metal that is placed on the ground to provide habitat for small animals. With our students we can use coverboards to practice monitoring the same area for life on a consistent basis. Under coverboards we can look for pill bugs, millipedes, worms, and slugs crawling through the rotting leaves. These creepy crawlies provide a great service to us. Just imagine how much dead plant material would pile up without them! In this activity, we are going to create new spaces where decomposers can live while making it easier for us to watch their work.

How

Decomposers like dark, cool, and moist places. Often this is underneath a rotting log. Since logs aren't always easy to lift, we are going to create a nice home that also allows us to easily keep tabs on them. First, you will need to find a cardboard box or piece of scrap lumber or metal to serve as your coverboard. If using a box, remove any tape and cut along one corner so that it can lie flat. If using a piece of lumber or plywood, make sure there are no nails sticking out. Next, find a place in your outdoor classroom or on the way to your space that does not receive a lot of heavy traffic, but with easy access. Students can even set coverboards out on the corner of the school building or parking lot, if there is soil and with permission. Finally, take a few dead leaves or other dead plant matter, like small twigs or grass, and place it underneath your cover board. If it is dry where you are, you may want to water the ground before placing your board on top. Come back the next day and see if any creatures have found the home you've made for them. It usually takes a day or two, but they should find it. Check it each day, and keep a record of what types of living creatures you find in a journal or with a camera, or create a checklist for species in your region. If you plan to have more than a few coverboards, number them and create a map of where they are, so you don't forget. This way you can keep data

Cover Board Survey

ECO
Educating Children
Outdoors

Time:_____ Date:_____
Class:_____ Site:_____
Weather:_____ Temp: _____

Animal	Picture	Tally		Notes
Worms				
Beetles				
Snails & Slugs				
Bugs				
Salamanders				
Spiders & Other Animals				

separate for each numbered coverboard you are monitoring.

If you live in an area with venomous snakes, insects, or spiders, use a snake hook or rake to lift the board. Lift the board toward you so there is a barrier between you and anything that has been residing underneath! Any time you replace the coverboard, be sure to do it slowly and gently. If you are doing this in a public area, be sure to remove your boards when you are finished with them.

As you're checking your coverboard, make note of what you find by asking: How many insects, invertebrates, and other small animals came to live under your coverboard? How many different types were there? Do you think you would have more or less if you tried it in a different spot? How have weather and change of seasons made a difference? If this is a lesson you continue year after year and keep accurate records, you may find some common trends of the inhabitants under the coverboard homes!

We Can Use These Materials Outside

- Coverboards, which can be made from cardboard boxes, scraps of lumber or plywood, or sheet metal
- Some dead leaves, small sticks, and other dead organic matter that can be placed under the coverboard
- Field guides of insects, amphibians, reptiles, and invertebrates
- Magnifying glasses

When We Get Back Inside

- Continue researching the role of decomposers in our environment. Understand which ones can be potentially harmful. These can include spiders, caterpillars, ants, and more.
- Composting worms are an excellent way to extend the learning inside. Get a worm composting bin and start composting leftovers from snacks and lunches!
- Mealworms are another great way to bring lessons of the life cycle right into the classroom. How about mealworms for a low-key classroom pet?
- You may consider other community science opportunities after practicing with coverboards. The Citizen Science Association connects people from a wide range of experiences around one shared purpose: advancing knowledge through research and monitoring done by, for, and with members of the public (citizenscience.org).

Books about Worms

Wiggling Worms at Work, by Wendy Pfeffer

A Worm Called Wallace: A Children's Book about Nature & Self Worth, by Jamie Rose

Wonderful Worms, by Linda Glaser and Loretta Krupinski

Buzzing with Questions: The Inquisitive Mind of Charles Henry Turner, by Janice N. Harrington

Lesson 23. Knowing Birds: How a Little Bird Taught Me

Narrative by Jenny Lyle

It is a spring day in May, and our classroom of preschoolers are preparing for their ECO day in the forest. In our morning circle we plan our time by recalling what we discovered last week and what we might find this week. Our ECO basecamp changes each week as

Count, count the magic number, is everybody here? ·
One, two, three, four, five
Six, seven, eight, nine, ten
Eleven, twelve, thirteen, fourteen!
Our magic number is fourteen today!
As we sing our song, children and

Nick Neddo

spring unfolds, and students are full of discoveries and questions.

There is a sense of freedom and ease as the weather is warmer, and preparing to go out is fast and effortless, compared with winter dressing here in the Northeast. Children are lined up to the door that exits our classroom directly to the outside. Without instruction they begin to sing: "Count, count the magic number. Count, count the magic number . . ." These singing voices are filled with excitement as we prepare to take our learning back outside!

Count, count the magic number!
Count, count the magic number!

adults are counting each student who is ready to go. Children practice counting, and teachers confirm the number of students present for the day. We stop and sing this catchy song at several spots along our journey, and after activities such as games and forest choice time. We are building a safety practice that helps us to take care of each other. It is simply part of our learning outdoors!

This spring we have had a visitor over the past several weeks, and the children can't wait to reconnect with this special winged friend. We first heard this bird as we sat at our meeting space where we gather as a class. We heard the call over and over again. We put on our deer ears and listened to the bird: "Teacher, teacher, TEACHer! Teacher, teacher, TEACHer!"

This bird's call rapidly rose, getting louder and louder. The call was so loud it interrupted our discussion at the gathering space. You couldn't help but hear the repetitive bird call over and over, sometimes eight to thirteen times, repeating, "Teacher, teacher, TEACHer!"

Initially we heard it with our ears, then we *saw* this small bird with its distinctive markings! While this bird foraged on the ground near us, we noticed the black and orange stripes on the crown of its head and a black speckled chest. Our place in the forest is a mix of deciduous and coniferous trees with a canopy so dense at times it feels like a room. White pine needles are a thick carpet mixed with birch, cherry, and maple leaves, perfect for a ground-feeding bird. With a field guide and help from other knowledgeable friends in our learning community, the children and teachers soon identified this bird. We discovered that our new friend was the Ovenbird!

Back inside the classroom we spent a few days listening to the Ovenbird call from the Cornell Lab of Ornithology website and YouTube video links. We learned about Ovenbird nests, habitats, and behavior. We also looked closely at Ovenbird photographs and painted our endearing feathered friend on large pieces of white paper. Teachers scaffolded this interest by adding a variety of books, videos, and materials as we continued to support this newfound interest and connection to our world.

Amy Butler

The enthusiasm for this learning experience continued to be fluid and swung from the inside to the outside over and over, as children built a connection with this special species. Outdoors, we made Ovenbird nests. We built our own nests at our sit spots in the forest. Ovenbirds create a small oven-like mud nest close to the ground and hidden. At the same time, we sculpted small clay Ovenbirds and painted them inside. Several preschoolers brought their small clay birds up to their nests and played with them during forest choice time. The connections grew stronger and helped guide the learning through the remainder of the year. During this thread of learning, the sounds of Ovenbirds, the connection to the bird itself, and the nesting materials all found naturally in our outdoor environment provided children the opportunity to delve deeper into their thoughts and ideas.

We were lucky enough to hear the Ovenbird several more times and watch it as our school year came to a close. The Ovenbird, a sweet, winged friend, had become part of our learning community.

Knowing Birds—How a Little Bird Taught Me

Grade Level: Pre-K–Third

Objectives

- Students will be able to name three to five local birds.
- Students will explore size, shape, behavior, and vocalizations of these birds.

We Are Exploring These Academic Standards

NGSS.K-LS1-1. Use observations to describe patterns of what plants and animals (including humans) need to survive.

> The preschooler was intent on catching a bird. She continued to sneak slowly.
>
> The area of the forest quieted, as it decompressed from the activity of children who had moved onto another area, and the bird flew closer. The child chose to sit on my lap and continue to watch.
>
> She whispered in my ear, "I want to hold it, I want to catch her and keep her."
>
> "I wonder how the bird would feel if you caught it?" I asked.
>
> "Probably scared. And sad."
>
> "Hmm," I replied. "I notice it's looking for food. The bird needs food to give its body energy." The child's solution: "She can stay in my coat, where it's warm, and I'll feed her."
>
> I replied, "That's very kind. Taking care of others is practicing kindness."
>
> As the preschooler walked closer, the Ovenbird continued to feed.
>
> Soon the other children gave the coyote call to come back to the fire circle, because our morning was coming to a close. The preschooler rejoined her group and shared her news with the other children. "I saw the teacher-teacher bird. She was nice."

NGSS.1-LS1-1. Use materials to design a solution to a human problem by mimicking how plants and/or animals use their external parts to help them survive, grow, and meet their needs.

NGSS.2-LS4-1. Make observations of plants and animals to compare the diversity of life in different habitats.

NGSS.3-LS4-3. Construct an argument with evidence that in a particular habitat some organisms can survive well, some survive less well, and some cannot survive at all.

What

Birds are calling us into awareness each time we come in contact with them. Their ability to fly and to sing is captivating enough, and yet there's so much more to explore. This lesson uses the sit spot method and a child's curiosity to drive explorations of the local birds that can be encountered on a daily basis. Using our senses and simple prompts, we can guide students into a deeper connection with birds and the place where we live—wherever we are.

How

Florence Merriam Bailey, one of the first women ornithologists, said, "The best way of all is to select a good place and sit there quietly . . . and see what will come." The easiest way to get to know a bird is to study whatever species are right outside your door.

Here are five ways to look, listen, and wonder at the birds you see every day.

1. **Find birds in your neighborhood.** Set the intention during one of your days outside to look for your bird neighbors. Who shares this neighborhood with us? Where do they like to hang out? Where can we go to be with the birds? This means *any* species of bird. Got pigeons? Great! House Sparrows? Great! An Ovenbird? Wonderful! Don't know the bird? Encourage students to give it a name themselves. All birds are good birds. Students can write down their discoveries in a journal or on a giant sticky note back in the classroom. For older students, consider mapping where you find these birds, document where they eat, where they roost, where they perch, or even a favorite puddle for bathing.

2. **Pick a group or solo sit spot near birds.** In unit 1, lesson 6, the sit spot was introduced. Use this approach to get to know your bird neighbors. Birds will become comfortable with your visits and will move back into a space when everyone is still and observant. Try a group sit where all the students sit quietly for a certain amount of time. Use these prompts to stimulate some thinking: Do you think you'll see or hear a bird first? What do you think the birds will be doing today? If there are no birds, where do you think they are and what are they doing?

Students can bring their journals and colored pencils to their sit spots. Invite them to draw what they see from where they are sitting. Students can mark where they are sitting with an X. Can you draw what is around you? Bring a little handful of birdseed to leave as a thank you to the birds in the neighborhood. Put up bird feeders near the sit spots.

Set out a bird buffet!

The great thing about setting up bird feeders in an accessible spot is that you create a common space, really an indoor sit spot, for everyone to observe birds from. Here are some tips:

- Check seasonal guidelines for your state for safe bird feeder use. In Vermont, we have to be careful of attracting bears, so our state guidelines are December 1 through April 1 for putting feeders out for birds.

- Keep it simple so that you can care for the feeders. One to three feeders will be enough to attract birds. Consider a window feeder, a hanging feeder, and a suet feeder. Seed choice includes black sunflower, thistle/niger seed, and suet/fat blocks. Be aware of birdseed with peanuts or made in a factory with peanuts, in case any students have allergies.

- Keep field guides, scrap paper, and colored pencils nearby to encourage research of birds the students notice. Students can count or tally birds and draw common feeder species.

- Spending time observing birds at the feeders may play a role in wellness for your students too. It's a place to take a brain break by connecting with nature and a calm place to let strong emotions dissipate. And it can become a social spot where kids can just enjoy birds together.

3. **Make a list of your usual suspects (a.k.a. the neighborhood birds).** Print out and laminate 8.5 × 11-inch images of your "usual suspects." (These are the bird neighbors that you'll most likely see.) This should be ten birds or fewer. Stick with the ones you and the students have discovered. Use them as flashcards, try sorting them by size from largest to smallest, trace the silhouettes of the birds with a dry-erase marker, and bring them outdoors with you wherever you go. Add new birds to the collection as the seasons change, then sort them by season. Make a giant list on the classroom wall of all the birds you've encountered together, along with the date you first met them.

4. **Draw sound maps and get to know bird vocalizations.** Together as a group, start by closing your eyes and just counting all the different sounds that you hear. Hold up a finger each time you hear a new sound. Try sorting sounds made by people, plants, and animals. From your sit spot with your journal, what sounds do you hear? How can we record them on the sit spot map? When children discover a new bird vocalization, try drawing what you hear, mimicking what you hear with your own voice, or even conducting what you hear with your hands. Use the free Merlin app to explore bird vocalizations and record and identify what you're hearing in your schoolyard.

5. **Document your scientific discoveries.** Capture your observations in a journal. Use the prompts "I notice, I wonder, It reminds me of" each time with

Amy Butler

students to help them observe and make connections. Students can draw, scribble, sketch, note, and write down questions. Back inside the classroom students can use field guides to further explore birds. Share and compare what you're noticing with others! Take a look at any of the nature journaling books by John Muir Laws (or his website johnmuirlaws.com) for more ideas.

We Can Use These Materials Outside

- Free birding apps for your phone, such as the Merlin Bird ID app, will help you identify birds by sight and sound
- Folding laminated field guides, like the ones by Waterford Press, are waterproof and ready to fit in your backpack (waterfordpress.com/Products/series/pocket-naturalist/vermont-birds/)
- An unlined nature journal in a waterproof bag with colored pencils and a pencil sharpener

We don't need to go to wild places to experience birds, we don't need to "go away" to experience birds or to experience nature. Every bird is miraculous, when we slow down to look at them and listen to their voices. Birds have a strong connection with place and noticing them strengthens and deepens our connection to place. They've chosen to be where they are because that place provides them with the things they need: food, water, shelter and a place to raise their young. What do we need as human beings? What is the same? What is different? Connecting with birds where we live can be as simple as setting an intention to notice birds.

—*Bridget Butler*

When We Get Back Inside

- Research common birds for your state. Look to the local Audubon Society or your state fish and wildlife department for information and resources.
- Check out the Cornell Lab of Ornithology's website All About Birds (allaboutbirds.org) to research birds and bird behavior.
- Bring field guides to regional birds into the classroom and have them accessible for quick referencing.

Books about Birds

Ruby's Birds, by Mya Thompson

Bird Count, by Susan Edwards Richmond

Birds, by Carme Lemniscates

Birdsong, by Julie Flett

Have You Heard the Nesting Bird? by Rita Gray and Kenard Pak

A Bird Will Soar, by Alison Green Myers

She Heard the Birds: The Story of Florence Merriam Bailey, by Andrea D'Aquino

Bird Boy, by Matthew Burgess and Shahrzad Maydani

The New Birder's Guide to Birds of North America, by Bill Thompson III

Stokes Beginner's Guide to Birds: Eastern/Western Region, by Donald and Lillian Stokes

Lesson 24. Catch and Release

By Ken Benton

Ice begins to melt as the days grow warmer. The ground thaws and gives way to soft mud that squishes beneath every step. Rivers and streams swell with the influx of melting snow and ice. Slowly, as the waters recede, trout become active and feed on freshly emerged stoneflies.

decided that for our next ECO session, we would embrace the season and catch our own "fish" to sort and measure.

Wanting a natural material, so as to avoid littering for the "release" part of our catch and release lesson, we peeled sheets of birch bark off a freshly fallen tree.

Nick Neddo

It is springtime in Vermont, and many students are gearing up to go fishing with their families.

One first- and second-grade class, in the midst of a unit on measurement, was using the familiar fishing concept of whether or not a fish is a "keeper" to practice measuring in one-inch increments. In the classroom, students placed measuring blocks on a printed picture of a fish. If it was over a certain length, they could write "keeper" next to it. If it was too small, they would have to throw it back. Building off this exercise, we

We used trout stencils of three different lengths to cut out enough trout for every student to have one. The morning of our ECO session, the classroom teacher had students examine and compare pictures of the three trout species present in the local rivers.

"This one has polka-dots all over it," said one first grader in reference to the photo of a brown trout.

After looking at a rainbow trout, another particularly observant student noted, "That's not a rainbow, it's just a pink stripe!"

catch them. Instruct the students to take the first fish they come across, even if it is not the one they colored.

Now that every student has "caught" a fish, inform students that their fish come in three sizes, and they must sort themselves into three groups by joining classmates with the same sized fish. Give students a couple of minutes to sort themselves based on comparing fish size, and then take out whatever measuring device they are familiar using (rulers, one inch measurement cubes, tape measure, etc.) to begin measuring their fish to see if their fish is actually the same size as others in their group.

There should now be three groups comprising the smallest fish, the medium-sized fish, and the largest fish.

Students will now take turns releasing their fish into the stream, starting with the group of smallest fish first. Position the two other groups along the bank downstream from the release point so that they can watch the fish as they flow by. If the class has built a replica of a stream, attach string to the cut-out fish and move them downstream by having students pull them along!

We Can Use These Materials Outside

- Tools for measuring, such as rulers, one inch measurement cubes, and tape measures
- Field guides or laminated pictures of fish, such as the laminated guides to freshwater fish produced by Waterford Press (waterfordpress.com)
- Yarn or string for making an impromptu fishing pole

When We Get Back Inside

- As an extension to this lesson, have students construct fish made out of natural materials. Use black or white pieces of material that are 12 × 12 feet for a backdrop. This can be done indoors or outdoors!
- Learning about fish is a gateway to learning about entomology (the study of insects) specifically related to fish and aquatic life. What types of insects do fish eat, and what are their life cycles?
- Understanding where fish find shelter and breed in water introduces the attributes of streams, lakes, and rivers, as well as landscapes that surround the bodies of water.
- When a body of water holds a healthy population of fish, it means the aquatic ecosystem is healthy too. Clean water supports macroinvertebrates, plant life, and mammals that live nearby. How do we know if an aquatic ecosystem is healthy? Research the indicators of a healthy body of water.

Books about Fish

Fish Everywhere, by Britta Teckentrup

Salmon Stream, by Carol Reed-Jones

Trout, Trout, Trout! A Fish Chant, by April Pulley Sayre

Over and Under the Pond, by Kate Messner and Christopher Silas Neal

The Little Black Fish, by Samad Behrangi and Azita Rassi (translator)

A Different Pond, by Bao Phi

Lesson 25. Adopt a Plant: Finding Friends and Family in the Plant World

Narrative by Roberta Melnick

I have been an educator for more than two decades. I have tried many things to build a community of learners. The human social dynamic in any classroom is an essential component to the learning environment. How we support the peer ecology of a classroom can either hinder or help blossom a student's social and aca-

is based on using effective listening and speaking skills among the class.

First, I share the personal birth story of each plant. Where they came from and how I have cared for them. Whether I started them from a cutting, or perhaps the plant was given to me already fully grown. Students take turns asking

Nick Neddo

demic development. Over the years I have discovered that learning about plants is an effective and dynamic experience for my students, and it has helped students connect not only to themselves but to their larger peer group as well.

In the beginning of the school year, I introduce my new group of learners to the plethora of plants in the classroom. These common houseplants have become an essential component of my toolbox for building a learning community. One of the pillars of this experience is that it

questions and begin to make personal connections. Students then engage in making predictions on the age of the plant and their species and how they may have gotten either their scientific plant name (characteristics) or their plant given name (personalized). This sparks conversations and immediately gets the class engaged in forming questions that satisfy their curiosity and brings out their most sensitive and caring side. It is often as spiritual and genuine as it gets when establishing connections with a new and vibrant group of students! It helps me to learn about what makes them who they are, and it also gives me insights into their vulnerabilities. I can see lots of their personality and characteristic traits within a few short days.

Next, I ask the students to consider adopting a plant. They fill out a questionnaire about why they are choosing their plant to adopt. They have to formulate reasons and apply themselves to be considered worthy of being an adoptive caretaker of the plant!

This past year one student adopted a plant and named them "Linda." Linda was growing in a purple pot, and this matched the student's newly dyed hair color. It was an immediate bonding moment. This student explained that they wanted to adopt Linda because it gave them a chance to express their grief about a family member who had committed suicide. They told me that every time they watered the plant it seemed like their uncle who passed was being remembered.

Another student adopted a spider plant and named them "Mr. Spidey." This student loved spiders. They quickly took to caring for the plant by pruning them so they could root the "brother and sister spideys" and give one to their mom. Again, this process of connection showed how important family was to this student.

What a unique way for me, as their teacher, to see their personalities shine through. I consider my plants to be class pets, of course, in every sense of the word. This experience inside the classroom with plants scaffolded a whole other world of learning outside. The appreciation, care, and connection

Roberta Melnick

to plants that was cultivated inside the classroom transferred seamlessly to the outdoors.

When a child has an encounter with a plant outdoors, I've noticed an inherently quiet approach that promotes purposeful movements (slow and observing versus running through the woods like it's recess time). I believe this is because of our prior experiences in the classroom with adopting plants. The students' approach to plants during ECO is an opportunity for a higher level of thinking and planning that is genuine and allows for the science of interdependence to be observed and synthesized by our students. This is not something that can be achieved without the experience of being in nature and getting up close and personal with plants. I've noticed that plants provide a platform for curiosity and inquiry that commonly generates a range of questions from students. These ideas and inquiries often stimulate an emotional response because of the simple fact that each plant is a living organism. Many children innately want to know what it takes for the plant to survive, and they will compare that to what it takes for them to survive and thrive. Many individual lessons are about the life cycle of plants, and categorizing of plants, which again connects important life-securing conversations about how to survive and thrive as well as about loss and suffering. These are tough concepts to genuinely access inside the classroom walls, and they have a profound effect on our learners. Leave it to the plants to help teach the concept of growth mindset, grit, and perseverance. I always ask my students to stop . . . listen . . . and see what the plant wants to teach them. This results in careful student observations of the plants and leads to lots of questions, too. Then, the teacher is able to help the learner determine the answers to their questions based on the stage of the plant's life cycle.

One of my students was a struggling learner who balanced different medicine regimes each day. This child lived in poverty and with mental illness. They had a difficult time fitting in at school. Peer to peer relationships did not come easily, and they lacked the friendships the other students were able to cultivate. On this particular day, we were continuing our learning with plants with a lesson that invited students to "discover" a plant and name them based on their characteristics. When this student wandered off in search of a plant to observe and name,

they had a determination and curiosity that led to an instantly deep relationship between themself and a flowering yellow plant. They immediately started talking to me about protecting this plant from everyone else, and that no one would ever be able to hurt them. They communicated that their plant belonged to a family, and they had to be protected. The student then cried for a bit as they went through the mental and emotional process of adopting and naming this plant. They then gathered up the whole class and let us all know that "Lemony" was now a part of their family and needed protection.

They caused quite a stir about this plant and had the attention of *all* their classmates. Still feeling very emotional, they led us through the upper field and then into the orchard to find this plant in full blossom in shades of yellow. The student informed everyone that this plant had lots of siblings, which no one was allowed to hurt. Everyone had to be nice to Lemony. Without a doubt, everyone agreed to never hurt Lemony. The whole class fully supported this student in protecting the entire Lemony family.

Two weeks later, Lemony's flower had passed, and they were wilting on the ground. Surprisingly, the student moved on easily. Our next lesson was about plant structure, and the students were invited to dig up a plant and learn about their roots. The student chose to dig up Lemony! Observations were made and a pencil sketch was done. This was a big, wondrous moment for this student, and it expanded their world through the genuine connection to the plant world. There were no boundaries during the inquiry process, and their peer group joined them in wonderment and awe. They were glowing as bright as his "Lemony" daffodil, both strong and vibrant—the student and the spirit of the plant.

Adopt a Plant: Finding Friends and Family in the Plant World

Grade Level: Pre-K–Third

Objectives

- Students will observe the same plant at different times during the year.
- Students will understand the life cycle of a plant.
- Students will practice sorting and classifying plants based on their physical characteristics and structure.

We Are Exploring These Academic Standards

NGSS.1-LS3-1. Make observations to construct an evidence-based account that young plants and animals are like, but not exactly like, their parents.

NGSS.2-LS4-1. Make observations of plants and animals to compare the diversity of life in different habitats.

NGSS.3-LS1-1. Develop models to describe that organisms have unique diverse life cycles but all have in common birth, growth, reproduction, and death.

What

Like all living things, plants go through different stages during their lifetime. And like all living things, plants are both vulnerable and responsive. Plants are all around us and affect us on a daily basis. We eat plants, use their fibers for clothing and shelter, extract their medicine, and they give us the oxygen we breathe.

Plants are able to use their own resources to grow and spread, produce seeds or shoots to reproduce, prepare to survive through the winter, and then grow new parts in the spring, or go through the process of death and decomposition. Plants can't move, there is no opportunity to escape if things aren't going well! They *must* survive and thrive in their ever-changing environment. At all times of the year, whether dead or alive, plants have interesting patterns of growth and interaction with other living things that students can observe and make connections to. And plants are everywhere! From urban landscapes to rural areas, plants have an amazing ability to grow in the cracks and crevices and along the edges of human impact and development.

How

Students will adopt a plant, observe it over time, and record how it changes through the season. In this lesson students will be exploring "phenology," which is the study of how living things

Kelsey LaPerle

change cyclically in relation to climate. As phenologists, students will be observing the changing natural phenomena of plant life and recording what they notice. Students will pick out a plant to adopt for a month, two months, or the remainder of the school year. Spring is a good time for this lesson as students have learned skills in observation, and this is when we see an explosion of plant growth as winter transitions to spring. When students choose a plant, they will end up choosing either an annual or a perennial. Annuals go through their whole life in just one year. Perennials can live for many years without flowering. And some plants do not flower at all (for example, mosses and ferns).

Be sure to consider how far away the adopted plant is from the school building. You may want to visit your plant more than once a week (actually, this is highly encouraged). Other considerations are the amount of traffic over the area with adopted plants and whether or not the area gets mowed. Adopted plants need to be safely accessible to all students. Know which plants are potential hazards in your area such as poison ivy or wild parsnip, which both cause contact dermatitis.

Once they have chosen their plants and introduced themselves to their plants, have students tie a piece of flagging tape with their name on the plant. This will help students find their plant later in the year. Back in the classroom, have each of the students make special flags with the plant's botanical, Indigenous, or student-made names on them. These flags not only mark and remind students of where their plants are, but special flags and labels will also bring in other curious passersby to see what is going on!

How can students get to know their newly adopted plant friend? Here are some suggestions for younger students:

- Students can draw or note observations of their plants during each visit and compare between visits.
- Record plant growth by taking pictures. Print and hang pictures in the classroom.

- Adults can scribe children's observations. This can be done one on one or as a group share. Write down all the observations and see how they change week to week!
- After a few weeks invite students to notice the plants around their adopted plant. What do they notice? Are all the plants the same?

Here are some suggestions for older students:

- Students will record observations each time they visit their plant. These observations can include:
  - The size of the plant and whether it has changed (height, width),
  - The presence or absence of leaves,
  - The presence or absence of buds,
  - The presence or absence of flowers, and
  - The presence or absence of fruit.

- Have students include metadata such as the date, temperature, and weather conditions
- Make a map locating everyone's plants. Include topography, shade, and human and animal traffic patterns.

Here are some questions for everyone:

- Has anything changed for your plant since we last visited? How so?
- What did you notice about your plant today? Are the leaves the same color?
- Has anything interacted with your plant since you last saw it? Is there evidence of insects or animals eating your plant?
- What do you think could happen to your plant over the next month?

As humans we must first pay attention. We must look beyond what is quickly observed to be fully aware of how plants support themselves and the other organisms with which we live, and how they transform their environment. Then, after careful, close observation, we must ask the right questions to learn from them about how to live with purpose, agency, and intention. And maybe we can take on some of these behaviors. Their lessons are ours for the learning.
—*Beronda L. Montgomery,*
Lessons from Plants

We Can Use These Materials Outside

- Botanical field guides
- Journals or note paper and a pencil or colored pencils
- Magnifying glasses or hand lenses
- Flagging tape to mark the newly adopted plant (back in the classroom, students can make a more permanent and personal flag for marking their adoptee)

When We Get Back Inside

- Students can write a description of their plant and the story of how they met the plant. Prompt them with questions such as, What did you find interesting about this plant? Why did you choose it? What do you think it will look like in two weeks, four weeks, three months?
- Students can learn more about the history of their plant. Is it indigenous to this region, or did it travel from far

way? Is it edible or used for medicine? Can it be used as a fiber or a dye?

- Adopting a plant is another opportunity for connecting with community science projects. The USA National Phenology Network brings together citizen scientists, government agencies, nonprofit groups, educators, and students of all ages to monitor the impacts of climate change on plants and animals in the United States (usanpn.org).
- An extension of this lesson would be for students to specifically choose a plant in the budding stage. Students can record their observations through Project Budburst (budburst.org), a website run by the Chicago Botanic Garden that collects phenological data.
- Since the school year here in North America is not in session during the summer, a time when plant growth and change is most abundant, we can consider supporting students to come back in September and check in on their adopted plant. What a great way to start the school year, by visiting an old friend!

Books about Plants

Day with Yayah, by Nicola I. Campbell

My First Book of Houseplants, by duopress labs

The Seedling That Didn't Grow, by Britta Teckentrup

Green on Green, by Dianne White

Afterword

*We delight in the beauty of the butterfly, but rarely admit
the changes it has gone through to achieve that beauty.*
—*Maya Angelou*

Writing this book has felt like many stages of instars. I think of *instar* as the term used to describe the more secretive stages of a life cycle of an insect. Instars are the pauses between molts that we don't see on the colored handout depicting a butterfly changing from egg to adult. The instar is the time between developmental stages, after the shedding (or rather absorbing) of skin, resulting in the change of color, patterns, and the growth of new body parts. They are the messy, mysterious, and sometimes relieving stages of molting from larva (or nymph) to adult maturity. These instars can go unnoticed by the human eye, but their advancement is both critical and completely dependent on environmental factors such as food source, habitat, and the health of the ecosystem for survival.

During the process of writing this book and with the success of each of my personal instars, the story molted into the more developed version of itself. Each chapter either became ready to sprout wings like a dragonfly or move into the complete dissolution of the chrysalis stage. One might say that authoring a book is a lonely process, and at times it most certainly is. And, within each instar, I was not alone. A healthy, diverse, and vibrant ecosystem of humans and

more than humans sheltered and fed me along the way. There are twenty-two teachers and artists who contributed to this book in the form of written words, photographs, and illustrations. When a larva achieves complete metamorphosis and survives to become a butterfly, it is a result of its interdependence on the environment in which it lives. This body of work is a collective ideation and is interconnected among the people, the land, and the many living beings that inhabit it. I am grateful for the opportunity to watch this book emerge into its full story.

This book started as a tiny egg of an idea that I have shared with many educators in Vermont and beyond. Teaching outdoors is something many of us had done early in our careers. Although, for me, the actual practice of learning *with* nature, and the development of a nature pedagogy, came much later in my career. It wasn't until I had been mentored by other veteran teachers that I was able to begin to understand all the stages needed to build a sustainable and effective model of nature-based education within a public school system. Without the help and shared vision of public school teachers, their principals and administrators, the North Branch Nature Center (NBNC),

Questions for growth and metamorphosis

- What does it mean to create an inclusive learning community in nature with all students and all adults?
- How does our time outdoors change our relationships with ourselves, each other, and the land we are engaging with?
- How are these experiences in nature supporting antiracist work and social and environmental justice?
- Are the learning routines we are practicing in nature serving all students? Why or why not?
- What does it mean to have a sense of belonging? How can we cultivate belonging and healthy attachment in nature?
- In what other ways can we re-localize and restore nature-based knowledge in our communities?

and families and community members, this book you hold would not have been possible.

Over many years, the teachers in this book have worked in partnership with NBNC to design a nature-based curriculum for their students that partnered with the landscapes and changing seasons of Vermont. By building on a model of collective impact and centering equity, the schools we have worked alongside have been able to create nature-based programs that are resilient and responsive to changing conditions of their world. Even though the contributors to this book are located in Vermont, today's children, whether in urban or rural landscapes, spend a majority of their time with screens or in adult-directed activities. In addition, the cumulative effects of climate change, a global pandemic, and structural racism have negatively impacted the health and wellness of our children and therefore our communities. In developing partnerships with schools focused on systemic change,

we must be intentional in understanding each school's culture and the teachers' experiences and comfort levels with being in nature, and the students as well. To teach and to learn with nature, to be outdoors with students and sharing the benefits of nature connection, is a practice in building relationships and healing generational trauma. Much as with teaching inside the classroom, trust is foundational to cultivating genuine relationships and is of the utmost importance to the safety and engagement of the students when teaching outdoors.

The time we spend in nature with our students expands our awareness of how we view what time spent outdoors can look like. Nature is not just found in national parks or with the newest outdoor clothing. To quote author Rahawa Haile, "The most important nature is the nature close by." The nature close by is the nature seen in the clouds overhead, the natural world we can see while drinking lemonade with a grandparent on a porch, or the nature we experience

Appendix 1: Template for Creating an ECO Schedule

The following table is a template for creating a schedule for your time spent learning with nature. The routines for learning outdoors are listed at the left, and teachers can use the spaces to the right to fill in the duration of each routine, the learning objective for each, a description of the activity happening during that segment of ECO, and any materials that should be on hand or other resources that would be helpful.

Routine	Duration (minutes)	Learning objective	Activity description	Resources + materials needed
Nature museum and morning circle				
Cooperative games				
Snack and story				
ECO lesson				
Forest choice				
Sit spot journaling				
Closing circle				

We acknowledge Jon Young, Ellen Haas, and Evan McGown for inspiring our use of the core routines through their book, *Coyote's Guide to Connecting with Nature.*

Appendix 2: Benefits of ECO—Assessment of Healthy Risk-Taking and Learning Outdoors

This form is to help document and guide outdoor learning so the experiences are meaningful and safe for students and teachers. This template can be used for planning an outdoor session ahead of time, as well as for reflection at the end. This form can also be used to communicate the benefits of learning outdoors with teachers, school administrators, and the larger community.

Location:	Activities:
Date and weather:	Class/Grade:
Site considerations and impact:	Site amendments and care:

Today students had various opportunities to learn, grow, and practice skills in the following areas:

Social/Emotional	Physical/Motor planning
Creative expression	Cognitive/Academic

Safety, logistics, and follow-up:

List here safety concerns and things that worked:

Problem-solve weather, clothing, transitions, etc.

How will you follow through on making changes?

Appendix 3: An Ecological Impact Assessment

Use this template to document the space you plan to visit regularly with your class. When visiting a site consistently over a period of time, it is our responsibility to become stewards of the land. We can do this by practicing Leave No Trace (LNT, see appendix 13), by cataloging plant and animal species, and by learning about the Indigenous people of the land. It helps to frame this assessment by asking: What will be the negative and positive impacts of our relationship with this place?

Name of site:	Location/Address:	Current owners of site:

Acknowledgment of Indigenous people's tribal lands:

Site map and photos:

Describe the past and present use of this site:

Use this section to catalog plant species in these four domains:

Ground layer (plants above the soil):

Understory (shrubs, vines, small trees):

Canopy (tree tops):

Animals (mammals, birds, insects, reptiles, and amphibians):

Surface rock and water sources:

Human impact (structures or evidence of use):

Appendix 4: Site Assessment and Hazard Mitigation Form

Before bringing students outdoors, a suitable site needs to be selected. Part of selecting a site is assessing any hazards that exist at the site. The forms and acronym shared below will be helpful when assessing outdoor classroom site.

The location of the site will be shared with school administrators and local fire and rescue groups, including a map and evacuation plan in the event of severe weather, a natural disaster, or other emergency (see appendix 5 on safety protocols for more details on this).

Name of assessor:	Site location:
Guest assessor (if any):	Approved by:
Season and condition of site:	Date:

Hazard	Who might be harmed and how?	Are there existing control measures?	Follow-up (by whom and completion date)

The acronym ARISE is helpful for identifying and mitigating hazards in an outdoor setting.

Awareness: Build awareness about the hazard and its possible implications.

Respond: In what ways will we respond to this hazard? Are there seasonal or cultural considerations in our response?

Isolate: Can this hazard be isolated to keep people safe? How will it be isolated?

Substitute: Is there another place we can learn and use as an outdoor learning space?

Eliminate: Can the hazard be completely eliminated? Do you need additional help to do it, or can it be done now?

This is an adaptation of the Hierarchy of Controls, a system used to minimize exposure to hazards.

Appendix 5: ECO Protocols for Safe Outings

When it comes to taking students to an outdoor classroom, safety is the first priority. Students' safety is a teacher's responsibility, supported by the school administration and the greater community.

There are several things to think about, both before heading outdoors and while you're there, to ensure students have a safe and secure outdoor learning environment. Having safety protocols in place ahead of time means things run smoothly, even during an emergency.

This section includes selecting a site, creating an evacuation plan, planning ahead for weather, making class lists and taking head counts, and being ready with first aid and medications. Presented here are several protocols for planning ahead of time as well as for handling emergencies in the outdoor classroom. These include a missing student, a rogue dog, an unidentified person, ticks, and the sudden onset of severe weather. Additional, separate appendixes provide more detailed protocols for fire safety, tool safety, and plant safety.

Ahead of Time

Selecting a Site

A careful site assessment should be done to select the right place for an outdoor classroom. Use the Hazard Mitigation Form in appendix 4.

Establishing an Evacuation Route

In the event of a community emergency or natural disaster, have an evacuation plan in place ahead of time. Follow the school's evacuation plans and consult with your local first responders and police department.

Sharing Your Location

School officials should have a map of the locations of regular learning areas and routes to get there quickly and efficiently. The map should include a winter plan and natural disaster plan for reaching these outdoor learning sites, as well as emergency evacuation plans. Contact your local police department and first responders to let them know where all the outdoor learning spaces are so they can respond efficiently during an emergency. Make sure all maps, plans, protocols, and routes are on file with school and emergency officials. Teachers should establish ahead of time their point person(s) at the school for all communication regarding plans, leaving and entering the building, and emergencies.

Preparing a Class List

Create a list of all the students in the class. This list can be brought on the outing and can include attendance for that day so teachers can have an accurate head count. This list can also be a place to note any specific medical issues or family situations of students that may be pertinent when dealing with an emergency, and this information should be shared with other adults who are responsible for the group.

Using Radios, Cell Phones, and Whistles

All teachers should have cell phones or radios to communicate back to the school in case of an emergency. All forms of communication need to be tested before the first session for signal strength and range. Every teacher should have all necessary contact numbers in case of emergency, including the local police number, park officials, the school principal, and the school nurse.

Additionally, all teachers must carry an emergency whistle. The internationally recognized distress signal from a whistle is three blasts.

Checking the Weather

Check the weather a few days before your outing in order to be prepared. Clothing for the day, your location, and the lessons that will be taught are contingent on the weather. Check again the night before and the morning of your outing. We know weather can change quickly. Take note of wind speed, temperature, and incoming weather systems. These are all changing variables that can be used for learning and to be better prepared. The National Weather Service is a great resource for accurate forecasting. Having a weather app on your phone—with good radar—is another safety measure.

Preparing a First Aid Kit

All teachers are responsible for carrying a school-supplied first aid kit. The school nurse is responsible for ensuring the teachers understand what is in the first aid kit and how to use it. The school nurse should supply an epi pen for students with allergies and additional first aid for any specific student needs.

Educating about Ticks

Educate parents and students about tick sweeps, which are quick visual checks for ticks on the *outside* of clothing, and about tick-borne diseases in your region. Ticks can be picked up any time kids head outside, including at the bus stop, on the playground, during outdoor physical education classes and on field trips. Back at home, parents and students should do a thorough check for ticks, including on the skin and hair.

It is the school's responsibility to communicate that ticks may be a part of your environment. When children return home from school, they or their parents need to be checking for ticks. Deer ticks, for example, are extremely small and are potential carriers of Lyme disease and other tick-borne diseases. To help educate parents, students, and the school community about ticks and tick-borne diseases, look for websites of state health departments and the Centers for Disease Control and Prevention to share. Additionally, dressing in long pants and long-sleeve shirts helps keep ticks off the skin.

Dressing for the Outdoors

All students should be prepared with proper clothing for heading outdoors, and appendix 7 covers this in more detail. In summary, teachers should collect these items for the classroom ahead of time, and/or notify families and children that they should be dressed for spending time outside. Depending on the location, this could include weatherproof outerwear, hats and mittens, waterproof boots, and extra socks in case one pair gets wet. Also, where ticks are a concern, students should be dressed in long pants and long-sleeve shirts to help prevent ticks from getting onto their skin.

That Day

Taking Attendance

Before students leave the school building, classroom teachers should make note of how many students are in the class on that particular day. Teachers can mark any absent students on their class list and take a current head count for the day, after accounting for any absences.

All teachers and volunteers are responsible for being able to see students *at all times*.

Establishing a Meeting Spot

Once your class is at the outdoor classroom, have the group decide on a central location for meeting up. This central location is to be used if students are separated from the group or if there is an emergency. This is a spot to do a head count or roll call before any emergency evacuations.

Taking Head Counts

Throughout the outing teachers are responsible for taking head counts, using the list created that day, which should account for any absences. This head count is especially important after and during games or other activities where students are spread out. Students must be in sight of an adult at all times.

Communicating with the School

Leaving the Building: Teachers should sign out of the building and communicate with school officials the exact location of where the class is going, how long the outing will be, and the estimated time of return. If your location changes, school officials should be immediately notified.

Returning to the Building: Teachers should plan to radio or call back to the school about their departure at the end of the outing, and they will need to get permission to re-enter the building. This should be done from a designated check point. This also applies when bringing a child back to the school for bathroom use or because of behavior or sickness, for early release, or for any other reason.

Emergency Situations

Missing Student

If a student is deemed missing, all students should be gathered at a central location that has been decided for the group that day. Another head count and roll call are done. Adults confer to be sure that they have not seen the child. The time is checked and recorded, and a call is placed to the school stating that a child is unaccounted for. While an adult occupies the other students with a story or stationary game, the other adults do a sweep of the area. After an initial sweep, if the child is not found, the time is recorded again, and the school should again be notified. At this point, school officials are responsible for calling the authorities.

First Aid and Injury

If a student or adult suffers an injury, teachers should confer and discuss the severity of the injury. While an adult administers first aid, the school should be notified of the incident. Teachers will need to communicate the nature of the injury to the school nurse, and the school nurse will help determine if the student needs to return to the building. All injuries that are administered to must be reported to the school nurse and documented.

Unidentified Person

If an unidentified person is in the area and is exhibiting behavior that is unsafe or suspicious, all adults in the group should be alerted. Students should be gathered back to the central location that was determined for that day. Suspicious behavior should be reported to the police if necessary and students removed from the area. If someone is speaking to a student who is not a part of the group, teachers should intervene to ensure the safety of the child.

Dogs

Students are not allowed to approach or pet dogs. Students should be reminded that dogs (leashed and unleashed) are to be left alone. If any students have an extreme fear of dogs, this should be noted ahead of time and shared with all adults in the group.

Ticks

Teachers will remind students during every outing to do a tick sweep, which is a simple body check on the *outside* of their clothes for ticks.

Severe Weather

While teachers should always check the weather before heading outdoors , severe weather can crop up quickly. Severe and unexpected weather may include high winds, thunder and lightning, heavy downpours of rain, or a rapid drop in temperature. Natural disasters such as earthquakes are also something to think about. In the case of severe weather, teachers should follow the evacuation route that was predetermined during the site assessment. Before evacuation, students should be gathered at their central location and a head count and roll call conducted. Teachers then should follow the evacuation route with a "safety sandwich" until arrival back at the school building. Teachers will radio or call back to the school about their departure.

Appendix 6: Backpack and Equipment List for Teachers and Students

Backpacks are absolutely the best way for teachers and students to carry their gear to the outdoor classroom. The best backpacks are ones with hip and chest belts, because these distribute a heavy load to other parts of the body, so the weight of a loaded backpack isn't all on the shoulders. Consider getting donated backpacks that can be used just for outings. These packs will get dirty, wet, and sat on.

Following are some details on selecting and using backpacks and a table that lists everything that should be in the teacher's and students' backpacks.

Backpacks for Teachers

Consider a pack with a built-in rainfly and outside zipped pockets for quick, easy access to supplies, especially for the first aid kits and radio or cell phone. Look for a pack with a loop at the top for hanging from a tree branch to keep it off the ground.

Backpacks for Students

Teach students how to put on a pack and how to keep the pack close to their backs by cinching down the straps. Backpacks should only carry what is essential for an ECO day.

ECO-Friendly Snack or Lunch

Snack and lunch are great opportunities to practice Leave No Trace and to learn how certain foods can provide energy and warmth. Snacks that are high in protein and healthy carbohydrates will fuel children through chilly days. Consider encouraging families and cafeteria staff to pack snacks in sealable containers to avoid spills of leftovers inside backpacks (for example, prepackaged yogurt can be substituted for slices of cheese and crackers). Juice boxes and containers of milk are not needed if each child has a water bottle. Warm tea is a special treat on a cold day, and for this it is a good idea to invest in reusable teacups. As always, remember the Leave No Trace principles, which are included in appendix 13. In short: Pack it in, pack it out!

In the teacher's backpack		In the student's backpack
First aid supplies from nurse	Dry tinder and matches**	ECO friendly snack and/or lunch
Epi pen (if needed)	Pencil/pen and sharpeners	Water bottle (2 for full day)
Attendance list, marked with who is absent	Observation/recording notebook	Teacup
Cell phone/radio—charged and warm	Snack for self	Science journal in resealable bag
Emergency whistle (these are often on backpack sternum straps)	Water bottle for self	Pencil and sharpener
	Extra water bottle for students	Extra layers of clothing**
Emergency contact numbers	Camera or video recorder	
Time piece (watch or phone)	Field guides**	
Toilet paper	Knife	
Rubber gloves	Scissors	
Small trowel	Folding saw**	
Hand sanitizer	Whittling peelers**	
Resealable bag for trash	Yarn/string	
Extra wool socks, mittens, hat, neck warmer**	50 feet parachute cord	
Laminated protocols	Picture or chapter book	
	Tape measure	
	Thermos with warm tea**	

**This depends on the season and activity.

Appendix 7: ECO Clothing List for Teachers and Students

Clothing that is appropriate for the weather is essential to having a productive and good time outdoors. During wet and winter weather, teachers and students need to be dressed in items made from materials that stay warm and dry in the elements. Layering is always a good idea, too. Parents and students need to be provided with clothing lists ahead of time. If families do not have the resources, you need to have these items available in the classroom for students. The table below lists essential clothing for teachers and students specific to the Northeastern United States. Use this template seasonally and regionally to list the gear and materials needed to support outdoor learning.

Teachers	Students
Wool or fleece hat that covers the ears Neck warmer Mittens/gloves Long underwear top and bottom Wool blend socks (and an extra pair) Rain jacket with hood Fleece or wool sweater Fleece or down vest Rain pants or bibs Winter coat with hood Snow pants that are water repellent Waterproof boots Winter boots that are lined	Base layer • Long underwear tops and bottoms • Wool blend socks Play layer • Fleece pants • Wool or fleece top • Neck warmer Outer layer • Waterproof winter coat with a hood • Waterproof rain/mud pants with bib or snow pants with a bib and snow skirt, plus reinforced knees and leg hems • Mittens that are insulated and waterproof • Boots with a wool liner and drawstring at the top • Wool or fleece hat that covers the ears

Appendix 8: Getting Gear for Outdoor Learning

It is our responsibility as school administrators and teachers to supply the needed clothing to students for outdoor learning. Having the appropriate gear to be outdoors in all weather is a barrier to many. In order for children to feel comfortable and safe, we need to outfit them in clothing that allows them to be an active participant when learning with nature.

First, before asking for funding or donations of clothing, be ready to explain your plan and the importance of learning outdoors with nature. Why are you teaching outside? Have your elevator speech (or trailside speech) ready. This will inform others about the goal of nature connection and how the community can be a part of it. Share your knowledge of this exciting project, and in doing that, you can also build alliances for support. Start in your school and your greater community first. Remember: long-lasting solutions that are resilient come from the interaction of many people!

Perform a needs assessment about clothing

- What is the weather of your region? Know the clothing you will need for all the weather of your region. How often and for how long will you be going out? This matters in layering for warmth and staying dry.
- Inform families as early as possible about the clothing needed. Create a gear list for your region (see appendix 6). What gear do parents and caregivers already have, and where are the gaps?
- Where will the gear be stored in your school? Is the clothing accessible and secure in closeable containers?
- How will the clothing be maintained and kept clean? Do you have a place to dry mittens and rain pants? Where can boots be organized and stored?
- Know that teachers and paraeducators need good gear, too!

Next, ask within your school

- Ask administrators what type of funding is available for students in need.
- Look to the school board, the parent teacher association, or a health and wellness committee that can help with a fundraiser or clothing drive. This also helps spread the word into the community about the benefits of learning outdoors
- Speak with the school nurse about extra clothing and securing resources for being outdoors for extended periods of time.
- What is left in the lost and found at your school or neighboring schools?
- Communicate with colleagues about the need for gear and whether a clothing donation box can be set up in a teacher/break room. This means clothes for students and adults!

Then, reach out into the greater community

- With your trailside speech ready and some photos of students already learning outdoors, reach out to local businesses for donations. Who will support this work? Local outdoor clothing stores, hardware stores, banks, and bigger corporations are all possibilities.

- Consider grants, matching grants, or seed money to help you purchase items.
- Are there corporate giving programs or other types of charitable funding in your community?
- Recruit community members to keep an eye out at thrift stores and yard sales for clothing. Some parents or caregivers might really like cruising online used market-places for gear!
- Start a seasonal clothing swap at the school for gear. This can be organized with the purpose of building an outdoor gear library.
- Ask community members who can sew to make neck warmers. Or better yet, find a volunteer who can come into the classroom and teach children how to sew their own!

Resources

Natural Start Forum: Removing Barriers to Outdoor Gear: https://naturalstart.org/bright-ideas/forum-outdoor-gear

A Forest Days Handbook: Program Design for School Days Outside, by Eliza Minnucci and Meghan Teachout

Appendix 9: Fire Safety Checklist

The following checklist is a quick and easy set of guidelines for having safe fires during outings. The list is designed to be photocopied, laminated, and brought in the teacher's backpack to the outdoor classroom. Appendix 10 goes into more detail and teaches the basics of starting, tending, and extinguishing a fire outdoors. Many of the points presented in this checklist will serve as reminders of the details explained in appendix 10.

Checking the Weather

- Check the weather using www.weather.gov.
- Check for wind, drought conditions, and fire bans.

Selecting a Site

- Consider roots, overhanging branches, ground litter, uneven ground, and runoff.
- Dig a fire pit and line the fire pit with sand or small gravel, or use a portable fire bowl.
- Build a fire ring of rocks around where the fire will be built with a clear boundary for students.

Collecting Materials

- Practice low impact harvesting and collect materials away from camp. Practice the Three Ds: dead, down, and dry. Do not collect firewood that has lichen or moss on it as these are living organisms we do not want to burn.

Practicing Fire Manners

- Teach students to use quiet voices and have quiet bodies during a fire.
- Students should be sitting only.
- When leaving the fire, students should walk around the outside of the safety boundary.

Tending the Fire

- Assign one adult tender and one student helper. The adult places sticks in the fire. Student helpers can sort, break, and hand sticks to the adult fire tender.
- The adult fire tender will model safe practice when tending the fire. Slow body, sitting position, attentiveness to fire, and following all fire manners.
- The fire tender will bring attention to when a fire is being lit. Only adults use matches. Once the fire is lit, the fire tender will place matches back into a waterproof container.
- All sticks that touch the fire go into the fire. The fire tender will decide a designated stick for fire stirring and moving the fire.
- The fire tender should wear fireproof gloves when handling cookware over the fire.

Extinguishing the Fire

- The fire tender will extinguish the fire at the end of the lesson with water. Snow and ice may also be used. Do not step on the extinguished fire. The student fire helper can be a part of this process.
- Say thank you and goodbye!

Appendix 10: Fire Building Basics

Selecting a Fire Site

The first step in building a fire is selecting and preparing a safe location to build the fire, and this will vary, depending on your location. In general, find a location that is relatively flat, clear of vegetation, and without low, overhanging branches that may burn or drop melted snow. The area should have ample room for students to move around the fire area safely. It should be free from roots, rocks, and uneven ground.

Preparing a Fire Site

Once a location is found, be sure to clear the surrounding area of debris that may ignite, such as sticks, leaves, or pine needles. Clear a circle at least five feet in diameter to remove any sort of ignition source outside the designated fire pit, so the fire does not spread.

After any flammable debris has been removed from the area, dig a shallow pit and create a ring of stones to contain the spread of fire and ash. Set aside the soil that was removed from digging the fire pit to be used if and when the fire pit is erased. If this fire area is intended to be used over many years and will be a permanent gathering space, you can line the fire pit with crushed stone or gravel. Be sure not to use stones from a river or stream bed, because the moisture that may be inside the stones turns to gas and expands as it is heated up, which may cause the rock to explode.

Once a clear fire ring has been established, set up a fire boundary with logs or larger stones. The only people allowed within this boundary are the fire tenders. Then, introduce students to proper fire safety procedures.

Building a Tinder Bundle

Now that the area has been cleared of flammable debris, a fire pit has been dug and surrounded with stones, and a fire boundary has been laid out, it is time to collect fuel for the fire. First, you need a tinder bundle that will catch the initial heat source (from either a match or a spark from striking flint and steel) and ignite to create a larger flame that in turn ignites the sticks that will be placed around the tinder bundle, described in the sections below.

To create a tinder bundle, use highly flammable material such as birch bark, cedar bark, processed jute twine, or dry grasses to create a small bundle resembling a bird or mouse nest. If using birch bark for a tinder bundle, be sure to tear the bark into thin strips and tease apart the layers of the bark. Birch bark is a particularly wonderful material for fire starting due to its high oil content, which allows the bark to ignite even when wet (a good thing to remember on a cold, rainy day). When using cedar bark, a similar technique is used where you pull the bark apart into strips as thin as possible. Once the cedar bark is torn into strips, vigorously rub the strips together in your hands, pausing occasionally to pull the bark apart even more.

Combining different materials into your tinder bundle often leads to better results.

understanding your students and bathroom habits is key. If a child should need to bathroom outdoors, they must always ask an adult first. Wild toileting outdoors should happen in a predetermined designated area. The adult present must be sure that the student has privacy. For solid human waste, dig a hole, often called a cathole, that is six to eight inches deep. To limit impact, toileting should take place at least two hundred feet away from any natural water source, basecamp, or trails. Each teacher should have an emergency toileting kit that includes disposable gloves, hand sanitizer, toilet paper, wipes, and gallon-size resealable bags for disposing of waste. Learn more about toileting outdoors in a nature-based program online at https://creativestarlearning.co.uk/early-years-outdoors/where-to-go-when-you-need -to-go/.

- If cooking is performed, any dishes or cookware should be washed either inside or two hundred feet away from any natural water source and with biodegradable soap. If washing cookware outside, dishwater should be either broadcasted out or, for larger amounts, a sump hole should be dug that is ten inches deep, allowed to percolate, and then filled in and covered.

Leave What You Find

- If you move a rock or log in search of insects, salamanders, etc. put it back after you are finished. These are homes to wildlife and should be respected.
- Sometimes activities call for using rocks, sticks, and other natural objects for building. When we do this, it is important that we only use non-living materials found on the ground. Living materials, such as moss, ferns, grasses, or branches still attached to the tree, should not be used.
- Old structures or structures that are no longer in use should be dismantled and restored to their natural setting. For example, the leaf litter and debris of a shelter should be spread out over a wide area so that it looks much like you found it. The same holds true for sticks and any other natural building material that may be used.
- Some activities may require the use of small amounts of living material. In these cases, the instructors will need to model appropriate harvesting practices.

Minimize Campfire Impacts

- Make use of Kelly Kettles, which are portable campfire stoves. They allow users to build small fires that leave zero impact.
- If possible, use preestablished fire rings rather than creating a new one. When creating a new fire site, be sure to clear the surrounding area of debris that may ignite, such as sticks, leaves, and pine needles. Dig a pit and create a ring of stones to contain the spread of fire and ash.
- Keep fires small and put on only what can burn completely in your time outside. Small fires use less material, put out less smoke, and are easier to manage.
- Use only dry material found on the ground to build fires. Harvesting green or living material damages trees and does not burn as well. Standing dead wood (such as dead trees that have not yet fallen over) should not be used as it provides food

and shelter for many animals and may be very dangerous to harvest. Be conscious of dead wood covered in lichen. Lichen not only holds moisture, it is also a living thing.

- Burn the fire down completely to ash, then put out with water. Do use soil to put out a fire as it may not put out all the coals. Also, do not stamp out a fire with your foot. (For more information on fire protocols and fire safety, see appendix 9).

Respect Wildlife

- If animals are encountered on the trail or in camp, observe them from a distance and respect their space.
- When students are allowed to handle small non-venomous animals, such as spiders, insects, frogs, or similar, it is important to be gentle and return them to where they were found. Teach students to hold living creatures with a quiet body and a quiet voice in a sitting or squatting position close to the ground.
- Never feed animals or leave food scraps outside.

Be Considerate of Other Visitors

- Other people outside your group may visit the same spaces as you do. It is important that we clean up after ourselves and leave our sites as we find them so that others can enjoy them.
- Loud noise not only disturbs wildlife, it also disturbs other visitors. Being aware of your noise level outside betters the experience for yourself and others.

The member-driven Leave No Trace Center for Outdoor Ethics teaches people how to enjoy the outdoors responsibly. This copyrighted information has been reprinted with permission from the Leave No Trace Center for Outdoor Ethics, which can be found online at www .lnt.org.

References

Banning, W. and Sullivan, G. 2011. *Lens on Outdoor Learning*. St. Paul, MN: Redleaf Press.

Brown, Maree Adrienne. 2017. *Emergent Strategy: Shaping Change, Changing Worlds*. Chico, CA: AK Press.

Charles, C. and Samples, B. 2004. *Coming Home: Community, Creativity and Consciousness*. Fawnskin, CA: Personhood Press.

Cobb, Edith. 1977. *The Ecology of Imagination in Childhood*. New York: Columbia University Press.

Comstock, Anna B. 1986. *Handbook of Nature Study*. Ithaca, NY: Cornell University Press.

Cornell, Joseph. 1989. *Sharing the Joy of Nature*. Nevada City, CA: Take Heart Publications.

Gans, Caylin. 2019. *Forest Schooled: The Book*. Caylin Gans.

Gill, Tim. 2010. "Nothing Ventured . . . Balancing Risks and Benefits in the Outdoors." London: English Outdoor Council.

Gill, Tim. 2007. *No Fear: Growing Up in a Risk Averse Society*. London: Calouste Gulbenkian Foundation.

Greenland, Susan K. 2016. *Mindful Games: Sharing Mindfulness and Meditation with Children, Teens and Families*. Boulder, CO: Shambala Publications.

Greenwood, D. A. 2012. Chapter 9. A Critical Theory of Place-Conscious Education. *International Handbook of Research on Environmental Education*. New York: Routledge.

Hanscom, Angela J. 2016. *Balanced and Barefoot*. Oakland, CA: New Harbinger Publications.

Himley, M. and Carini, P. 2000. *From Another Angle: Children's Strengths and School Standards*. New York: Teachers College Press, Columbia University.

Holland, Mary. 2019. *Naturally Curious: A Photographic Field Guide and Month-by-Month Journey through the Fields, Woods, and Marshes of New England*. North Pomfret, VT: Trafalgar Square Books.

Knight, Sarah. 2016. *Forest School in Practice for All Ages*. London: Sage Publications.

Knight, Sarah. 2011. *Risk and Adventure in Early Years Outdoor Play*. London: Sage Publications.

Knight, Sarah. 2011. *Forest Schools for All*. London: Sage Publications.

Larimore, Rachel A. 2019. *Preschool Beyond Walls: Blending Early Childhood Education and Nature-Based Learning*. Lewisville, NC: Gryphon House.

Louv, Richard. 2011. *The Nature Principle: Human Restoration and the End of Nature Deficit Disorder*. Chapel Hill, NC: Algonquin Books of Chapel Hill.

Louv, Richard. 2005. *Last Child in the Woods: Saving Our Children from Nature-Deficit Disorder*. Chapel Hill, NC: Algonquin Books of Chapel Hill.

Martin, Peter. 2007. "Caring for the Environment: Challenges from Notions of Caring." *Australian Journal of Environmental Education* 23: 57–64.

Minnucci, E. and Teachout, M. 2018. *A Forest Day Handbook: Program Design for School Days Outside*. West Brattleboro, VT: Greenwriters Press.

Montgomery, Beronda L. 2021. *Lessons from Plants*. Cambridge, MA: Harvard University Press.

Natural Start Alliance and Merrick, Christy, ed. 2019. *Nature-Based Preschool Professional Practice Guidebook*. Washington, DC: North American Association for Environmental Education.

Palmer, Parker J. 1998. *The Courage to Teach: Exploring the Inner Landscape of a Teacher's Life*. San Francisco: Jossey-Bass.

Robertson, Juliet. 2017. *Messy Maths: A Playful, Outdoor Approach for Early Years*. Carmarthen, UK: Independent Thinking Press.

Robertson, Juliet. 2014. *Dirty Teaching: A Beginner's Guide to Learning Outdoors*. Carmarthen, UK: Independent Thinking Press.

Sobel, David. 2015. *Nature Preschools and Forest Kindergartens: The Handbook for Outdoor Learning*. St. Paul, MN: Redleaf Press.

Sobel, David. 2008. *Childhood and Nature: Design Principles for Educators*. Portland, ME: Stenhouse Publishing.

Sobel, David. 2002. *Children's Special Places: Exploring the Role of Forts, Dens, and Bush Houses in Middle Childhood*. Detroit, MI: Wayne State University Press.

Sobel, David. 1996. *Beyond Ecophobia: Reclaiming the Heart in Nature Education*. Great Barrington, MA: The Orion Society.

Topa, Wahinke (Four Arrows) and Naravez, Darcia. 2022. *Restoring the Kinship Worldview: Indigenous Voices Introduce 28 Precepts for Rebalancing Life on Planet Earth*. Berkeley, CA: North Atlantic Books.

Wall-Kimmerer, Robin. 2013. *Braiding Sweetgrass: Indigenous Wisdom, Scientific Knowledge, and the Teaching of Plants*. Minneapolis, MN: Milkweed Editions.

Warden, Claire. 2015. *Learning with Nature: Embedding Outdoor Practice*. London: Sage Publications.

Warden, Claire. 2012. *Fascination of Earth: Wood Whittling*. Auchterarder, Scotland: Mindstretchers.

Warden, Claire. 2012. *Fascination of Fire: Charcoal*. Auchterarder, Scotland: Mindstretchers.

Williams, Florence. 2017. *The Nature Fix: Why Nature Makes Us Happier, Healthier and More Creative*. New York: W.W. Norton.

Young, J., Haas, E. and McGown, E. 2010. *Coyote's Guide to Connecting with Nature*. 2nd ed. Santa Cruz, CA: OWLink Media.

Young, Jon. 2018. *What the Robin Knows: How Birds Reveal the Secrets of the Natural World*. Boston: Mariner Books.

Contributors

Narratives, stories, lesson plans, photographs, illustrations, and a love for the natural world were contributed by the following teachers. Many thanks to them for their time and dedication in helping create the story of ECO!

Angie Barger is a clinical herbalist and mindfulness educational consultant in Marshfield, Vermont. She is the founder of the Tea Project, a virtual series of courses for educators founded on the principles of nourishing teachers, students, and school communities.

Ken Benton has had the honor of helping children form their own outdoor memories as a North Branch Nature Center teacher-naturalist for the past ten years. He's helped bring ECO to many public schools through-out central Vermont, in addition to running camps, after-school programs, and other outdoor programming. While Ken is not birding, hunting, or fishing with his son, he can be found foraging and creating delicious meals harvested from the wild.

Ainsley Burroughs is a kindergarten and first-grade teacher at Barre Town Elementary School in Barre, Vermont. She loves adventuring and learning from the natural world alongside students in their outdoor classroom.

Bridget Butler, also known as the Bird Diva, teaches a mindful birding practice called Slow Birding that connects people, birds, and place. She leads courses and workshops on birding both online and in-person. She's also a bit of a bird whisperer and is well known for her haunting owl calls!

Emily Carley, M.Ed., is a literacy specialist and kindergarten and first-grade teacher at Union Elementary School in Montpelier, Vermont. When she's not teaching or creating lessons to help her students become happy readers and writers, you can find her seeking her own artistic inspiration from nature. Emily's quick to take watercolors out of her pack to paint dawns and dusks, venture off to sift through tidepools at the ocean, and always eager to cook over an open fire.

Whitney Doenges teaches first and second grade at the Warren Elementary School in Warren, Vermont. She loves being outdoors with her students, teaching, writing, and fostering independence and creativity in the classroom and beyond.

Pam Dow teaches at the Moretown Elementary School, in Moretown, Vermont, where she has been learning beside young children in grades pre-K through second for more than twenty-five years. She regularly takes her students into the woods behind the school to explore, create, question, and learn with nature.

Gina Gaidys teaches first and second grade at the Warren Elementary School in Warren, Vermont. She loves supporting students to grow their powers of persistence, resilience, and empathy, both in the classroom and in the natural world.

Leah Greenberg loves taking time to notice, wonder, and play—both with her students and with her camera. From San Francisco, where she lived for twenty years, to Vermont, where she currently teaches kindergarten, Leah is dedicated to learning about and caring for the people and places she encounters.

Harriet Hart teaches preschool at Braintree Elementary School in Braintree, Vermont. They take great delight in creating stories and songs to support their students' exploration of local flora and fauna. Harriet occasionally leads storytelling workshops for early childhood educators.

Brenda Hartshorn, a retired kindergarten through second-grade teacher from Moretown Elementary School in Moretown, Vermont, dedicated the last ten years of her thirty-eight-year teaching career to educating children outdoors. She found ways to implement math, literacy, science, and social emotional learning in the woods, with nature being the lead teacher. Brenda has been a coteacher for the ECO Institute through North Branch Nature Center, alongside Amy Butler, supporting teachers who strive to bring their students outside the classroom walls to engage with authentic, meaningful learning.

Pete Kerby-Miller is a teacher-naturalist at North Branch Nature Center in Montpelier, Vermont, where they work with public school teachers to educate children outdoors. They strive to bring joy, empowerment, and justice into every lesson.

Susan Koch is a lifelong learner and a first-grade teacher at Union Elementary School in Montpelier, Vermont. She believes that developing a sense of wonder is crucial for the future stewards of the Earth. Susan can be found in the forest each Thursday with a Kelly Kettle full of hot tea and her binoculars.

Kelsey LaPerle is a preschool teacher at Barre Town Middle & Elementary School who loves experiencing nature alongside her students. Kelsey especially loves immersing her students in nature for the first time in their lives and witnessing the wonder and awe of their young minds. When she is not teaching, Kelsey can be found hiking, biking, paddling, or otherwise adventuring outside.

Jenny Lyle teaches preschool at Moretown Elementary School in Moretown, Vermont. She loves building community, learning alongside young children, and exploring the natural world.

Roberta Melnick works in Barre City as the third- and fourth-grade English language arts and social studies looping teacher. Her passion is igniting students through the nature of taking risks and challenging themselves to make each day count. Being immersed in the natural world, she feels, supports a genuine experience that enriches our lives together as a learning community.

Fiona Modrak grew up in upstate New York, where she learned a love of hiking and the outdoors from her family. After graduating with a bachelor's degree from Cornell University, she spent several years teaching science curriculum outdoors across the country, before earning her master's degree from the State University of New York's School of Environmental Science and Forestry. She now works in Vermont, bringing students outside to learn about science and the natural world.

Dave "Muskrat" Muska is a naturalist and educator at the North Branch Nature Center in Montpelier, Vermont, where he has worked outdoors with children for more than a decade. He is also the founder of Ondatra Adventures, a practice devoted to inspiring and developing personal relationships with the natural world through naturalist study and outdoor excursions.

Nick Neddo understands the heart and soul of the work this book represents. The artwork that helps to bring this book to life is coming from Neddo's hands and heart, through more than twenty years of outdoor education experience. His commitment to teaching people about the natural world, and helping people rediscover their kinship with the landscape, is as much of a focus for him as is making artwork. He is the author of *The Organic Artist*, and *The Organic Artist for Kids* (Quarry).

Jenna Plouffe facilitates Robin's Nest Nature Playgroup and teaches preschool at North Branch Nature Center's Forest Preschool in Montpelier, Vermont. She finds joy in exploring and being curious about the natural world with all the children in her life. She loves witnessing small but significant moments of nature connection.

Carrie Riker returned to North Branch Nature Center in 2014 as a teacher after a long hiatus. Her love of nature began with making mud soup, onion grass pancakes, and dirt cookies in her backyard as a child. This love has continued to grow throughout time as she enjoys sharing big and small nature moments with youth of all ages at NBNC.

Jillian Zeilenga is a kindergarten and first-grade teacher at East Montpelier Elementary School in East Montpelier, Vermont. She loves exploring nature with her students in their outdoor classroom and jumping in puddles with them.